Pvt. Ltd.

Promotion of
Public - Bred Hybrids of Field Crops

S.K. Rao
Professor and Head
Plant Breeding and Genetics
JNKVV, Jabalpur - 482 004

G. Kalloo
Vice Chancellor
JNKVV, Jabalpur - 482 004

D. Khare
Senior Scientist
Plant Breeding and Genetics
JNKVV, Jabalpur - 482 004

2010

Studium Press (India) Pvt. Ltd.

Promotion of

Public - Bred Hybrids of Field Crops

ISBN: 978-93-80012-04-9

Published by:

Studium Press (India) Pvt. Ltd.
4735/22, 2nd Floor, Prakash Deep Building
(Near Delhi Medical Association),
Ansari Road, Darya Ganj, New Delhi-110 002
Tel.: 23240257, 65150447; Fax: 91-11-23240273;
jngovil@gmail.com; jngovil@hotmail.com

Printed at:

Salasar Imaging Systems, Delhi-35

CONTRIBUTORS

Chatrath Ravish: Directorate of Wheat Research, Karnal - 132 001.

Chowdary P.N.: Vikky's Agri-sciences Pvt. Ltd., Hyderabad - 500 034.

Dhua S.R.: Crop Improvement Division, Central Rice Research Institute, Cuttack - 753 006.

Elangovan M.: National Research Centre for Sorghum, Rajendranagar, Hyderabad - 500 030.

Gupta H.S.: Vivekananda Parvatiya Krishi Anusandhan Sansthan (Indian Council of Agricultural Research), Almora - 263 601.

Hegde D.M.: Directorate of Oilseeds Research, Hyderabad - 500 030.

Hora Manpreet: Dept. Plant Breeding and Genetics, JNKVV, Jabalpur - 482 004.

Kalloo G.: Vice Chancellor, JNKVV, Jabalpur - 482 004.

Khairwal I.S.: AICPMIP Mandor, Jodhpur - 342 304.

Khare D.: Dept. Plant Breeding and Genetics, JNKVV, Jabalpur - 482 004.

Kumari Sangeeta: AICPMIP Mandor, Jodhpur - 342 304.

Mahajan V.: Vivekananda Parvatiya Krishi Anusandhan Sansthan, Almora - 263 601.

Mishra B.: Directorate of Wheat Research, Karnal - 132 001.

Mishra Rajesh: GM (R&D) Seeds, Dhanuka Agritech Ltd., Hyderabad.

Pandey M.P.: Crop Improvement Division, Central Rice Research Institute Cuttack - 753 006.

Prabakaran A.J.: Directorate of Oilseeds Research, Hyderabad - 500 030.

Rao B. Dayakar: National Research Centre for Sorghum, Rajendranagar, Hyderabad - 500 030.

Rao S.K.: Professor and Head, Dept. Plant Breeding and Genetics JNKVV, Jabalpur - 482 004.

Rao Y. Yogeswara: Vikky's Agri-sciences Pvt. Ltd., Hyderabad - 500 034.

Ravikumar S.: National Research Centre for Sorghum, Rajendranagar, Hyderabad - 500 030.

Ravinder Reddy, Ch: International Crops Research Institute for the Semi-Arid Tropics, Patancheru-502 324

Sain Dass: Director, Directorate of Maize Research, ICAR, New Delhi - 110 012.

Saxena K.B.: Principal Scientist, International Crops Research Institute for the Semi-Arid Tropics, Patancheru - 502 324.

Seetharama N.: National Research Centre for Sorghum, Rajendranagar, Hyderabad - 500 030.

Singh S.K.: Directorate of Wheat Research, Karnal - 132 001.

Subbian P.: Director, Directorate of Agribusiness Development, Tamil Nadu Agricultural University, Coimbatore - 641 003.

Tonapi Vilas A.: National Research Centre for Sorghum, Rajendranagar, Hyderabad - 500 030.

ABOUT THE AUTHORS

S. Koteshwara Rao University Professor & Head, Department of Plant Breeding & Genetics, Jawaharlal Nehru Krishi Vishwa Vidyalaya, Jabalpur (MP) has worked more than 29 years in crop improvement and developed several crop varieties. In recognition of his meritorious and out standing contribution in the field of Maintenance breeding, seed production and seed systems Dr. Rao has been conferred prestigious Rafi Ahmed Kidwai Award. Dr. Rao has published more than 200 research papers in international and national Journals. He is actively involved in planning and promotion of public sector hybrid technologies of Rice, Maize, Pigeonpea etc. in Madhya Pradesh as well as other parts of the country.

Gautam Kalloo Vice Chancellor, Jawaharlal Nehru Krishi Vishwa Vidyalaya, Jabalpur (M.P.) served as Deputy Director General (Horticulture), Deputy Director General (Crop Science), ICAR, New Delhi and Director, Indian Institute of Vegetable Research, Varanasi. He served 22 years in different capacities at CCS HAU, Hisar. He did his Ph.D. in Genetics and Plant Breeding and has been awarded D.Sc. *(Honoris Causa)* by Banaras Hindu University, Varanasi. He did his Post Doctorate from Germany.

He is an author/editor of several books published from Paragon Press Oxford, Elsevier - Netherlands, Springer-Verlag, New York/Germany, CRC Press, Florida the United States etc. and has over 250 research papers/ articles to his credit. He has developed more than 40 crop varieties and associated in developing a series of network projects including mega seed project.

In recognition of his meritorious and outstanding contribution in the field of Horticulture, Prof. Kalloo has been conferred upon with several prestigious awards including Rafi Ahmed Kidwai Award, ICAR Award of outstanding Multidisciplinary Team Research in Agriculture and Allied Sciences etc. He is a Fellow of National Academy of Agricultural Sciences besides President of Indian Society of Vegetable Sciences; President, Indian Society of Seed Science and Technology and; President, Indian Society of Oilseeds. He has widely travelled in various capacities. His mission is to modernize Agricultural Research System to enhance the efficiency and to generate' better technologies for the farmers and to establish best facility for teaching to produce quality students to serve the state/country.

Dhirendra Khare, Ph.D. is a Senior Scientist, Plant Breeding and Genetics at JN Agricultural University (JNKVV) Jabalpur, MP, India. He has about 80 publications, three edited and two text books apart from five technical bulletins published by ICAR on characterization of crop varieties, has 24 years teaching and research experience in plant breeding and seed technology. He is life and fellow member of Indian Society of Genetics and Plant Breeding and Indian Society of Seed Technology. At present, he is associated with Seed Technology Research of NSP (Crops) with main emphasis on characterization of crop varieties, along with research on pigeonpea hybrids and HQP maize.

PREFACE

The low productivity of crops in India remains a cause of concern to the policy makers, farming community and other stakeholders. The increase in agricultural production, however, has been brought, uneven development across regions, crops and sections of farming community. Government of India has focused on infrastructure development in the field of Agriculture. The expenditure can only be exploited by growing the management responsive varieties.

Development of technology and its access to people are two mutually exclusive dimensions of development. Achieving higher yield by hybrid vigour is a proven technology for vertical increase in agricultural production. Hybrid technology is exploited commercially in maize, sorghum, pearl millet, sunflower, cotton, pigeonpea, rice, castor, safflower, mustard etc. whereas in many crops the research is at advanced stage. However, a lot more attention is required to harness their immense potential in different crops.

Scientists from pubic sector had done meticulous efforts to exploit hybrid vigour in enhancing yield potential since 1960's with tremendous financial support of the Government. With the introduction of New Seed Policy in 1988, medium and large seed companies and companies with foreign collaboration played the major role in making the hybrid varieties and its seed available to the Indian farmer with the use of internationally available relevant technologies. Since then the genuine efforts of public sector in this field are not able to reach to the farmer upto acceptable frontier.

It is felt that in spite of the distinct advantages of hybrid technology, the area coverage is not sufficient for further increasing the productivity of major cereal crops. An organized seed industry, trained manpower and appropriate national policies to support public bred hybrid technology, for both research and development, are yet to be intensified.

Hybrids' proportion of total seed produced is not very satisfactory (cotton 23.0%, maize 59.6%, sorghum 77.6%, castor 75.0%, pearl millet 60.0% and sunflower 29.6%). The public sector hybrids are not getting place in market share as compared to private hybrids in spite of their potential yields. The achievements of genetic improvement of hybrid research of NARS have not been fully translated into increased productivity at the farm level to bring more income to poor farmers. In spite of massive Breeder and Foundation seed production programs undertaken by the SAUs and public sector agencies, the supply of certified seed of hybrids to farmers is insufficient, thus force them to grow farmers saved seed of age-old varieties with low yield potential.

A shift to sustainable agriculture production requires fundamental changes to the hybrid seed production and security systems with the development of hybrids in many field crops to utilize the expenditure made in the field of hybrid development by the Government.

JN Krishi Vishwa Vidyalaya, Jabalpur organized a National Seminar on **Strategies for commercialization of public bred hybrids in India** on February 23-24, 2008 to gather together research workers of both private and public sectors to contribute in the problem of national interest and to increase the service rendered by hybrid vigour in expanding food supply. The objective of the seminar was to develop farmer-preferred, improved hybrids of crops; knowledge empowerment of seed producer in hybrid seed production; to strengthen seed production chain in public sector seed system; to establish linkage between informal and formal seed sectors; research backstopping for further improvement of hybrids for resistance to biotic and abiotic stresses; to suggest policy and institutional changes, to ensure supply of quality seed of hybrids; and to stabilize the local seed supply and diffusion systems of improved hybrids.

The views of the renowned scientists actively engaged in exploitation of hybrid viogur in respective crops covering above objectives are presented in this book entitled **Promotion of Public - Bred Hybrids of Field Crops** for proper exploitation of the phenomenon by both the sectors to feed burgeoning world population.

We express our sincere gratitude to the authors contributing their views for promotion of public bred hybrids and to the University for the generous financial support. We also thank all the faculty and staff members of the University for their support for grand success of the conference.

We hope the omnibus will be useful for researchers and policy makers affianced in various agricultural pursuits.

Editors

CONTENTS

Contributors v

Preface ix

1. Promotion and Commercialization of Public - Bred Hybrids in India - An Overview 1
S.K. Rao, G. Kalloo and *D. Khare*

2. Strategies for Commercialization of Public - Bred Hybrids of Rice in India 13
M.P. Pandey and *S.R. Dhua*

3. Single Cross Hybrid Seed Production in Maize 41
Sain Dass

4. Strategies for Commercialization of Public - Bred Hybrids of Maize in India 51
H.S. Gupta and *V. Mahajan*

5. Strategies for Commercialisation of Indian Sorghum Hybrids Bred in Public Sector 67
N. Seetharama, Vilas A. Tonapi, B. Dayakar Rao, S. Ravikumar, Ravinder Reddy, Ch. and *M. Elangovan*

6. Strategies for Promotion of Public - Bred Pearl Millet Hybrids in India 107
I.S. Khairwal and *Sangeeta Kumari*

7. Strategies for Commercialization of Public - Bred Hybrids of Castor, Sunflower and Safflower in India 139
A.J. Prabakaran and *D.M. Hegde*

8. Breeding Hybrids for Enhancing Productivity in Pigeonpea 115
K.B. Saxena

9. Strategies for Commercialization of Public - Bred Hybrids of Wheat in India 165
S.K. Singh, Ravish Chatrath and *B. Mishra*

10. Promotion of Partnerships for Commercialization of Hybrid Technology 197

S.K. Rao

11. Agribusiness in India: Opportunities and Prospects 203

P. Subbian

12. Strategies for Commercialization of Public - Bred Hybrids Through Private Sector Partnership in India 209

P.N. Chowdary and *Y. Yogeswara Rao*

13. Hybrid Seed Production of Emerging Field Crops - An Overview 219

S.K. Rao and *Manpreet Hora*

14. Public - Private Sector Partnership 245

Rajesh Mishra

Index 253

1 Promotion and Commercialization of Public - Bred Hybrids in India—An Overview

S.K. Rao, G. Kalloo and D. Khare*

ABSTRACT

The responsibility to develop hybrids for different ecosystems and less profitable crops is owned by public sector. Vision is required to share responsibilities between private and public sectors, as in future public sector alone may not be able to face this challenge. At the same time, the relevance of public sector and up coming private sector requires continued working in crops / situations of interest of private sector as well. Desirable results require joint efforts. At national level, public sector seed corporations have not made significant impact in the field of hybrid seed production. This being a skill requiring procedure compared to inbred varieties, requires sensitization of these corporations, through planned efforts in human resource development besides policy initiatives encouraging hybrid seed production and popularization. Healthy competition between private and public sector should be rewarding. Issues of subsidy and incentive pricing for hybrid seeds are worth examining in view of anticipated benefits.

Key words: Alternative areas, commercialization, emerging issues, environmental implications, linkages, public bred hybrid, seed requirements

Seed is the carrier of the technological improvement ensuing from plant breeding research, a recognized means to improve productivity by genetic alteration. In the early era, genetic improvement for crop productivity was largely focused by public sector with only a supportive role from the private sector. The Government of India recognizing the need for continuous upgradation in seed/planting material introduced a new policy on seed development in October 1988. This policy enabled the medium and large seed companies and companies with foreign collaboration to bring the internationally available relevant technology to the Indian farmer. At this point of time, the advances in biotechnology and their integration with

* J.N. Krishi Vishwa Vidyalaya, Jabalpur - 482 004.

plant breeding is at an exciting phase of scientific development that calls for new policy initiatives to ensure absorption of the new technology in our country.

Seed is the most crucial, critical and vital input for attaining sustained growth in agricultural production. It is also proved, that hybrid seeds lead to vertical increase in agricultural production because of their much higher yield potential compared to high yielding open-pollinated varieties. However, a lot more attention is required to harness their immense potential in different crops.

Sufficient number of commercially viable hybrids, suited to different agro-climatic conditions, in crops like sorghum, pearlmillet, cotton, castor and also up to some extent in maize have been developed in India. The country is still far behind in exploiting heterosis commercially in rice, oilseeds and many vegetable crops. Number of public and private sector seed companies is operating for production of adequate quantities of quality seed, especially of hybrids in many crops, but still the quantity produced is far behind the actual requirement.

The public sector seed corporations have to play a pivotal role in producing and making available hybrid seeds to the farmers, who can derive benefit from new technology and meet the food requirements of the expanding population by way of increased production.

It is felt that in spite of the distinct advantages of hybrid technology the area, coverage is not sufficient for further increasing the productivity of major cereal crops,. An organized seed industry, trained manpower and appropriate national policies to support public bred hybrid technology, for both research and development, are yet to be intensified.

International organizations, *viz.*, ICRISAT, IRRI and CIMMYT are playing an important role in research on hybrid technology by way of exchange of useful germplasm, source of male sterility, and inbreds and restorers along with human resource development in hybrid technology.

The recent policy of these centres concerning Intellectual Property Rights for hybrids has also been well received both by the public and private sectors. IRRI has recently evolved a policy concerning hybrid rice, which states:

- IRRI adheres to the policy of free availability of the breeding lines, elite germplasm and parental lines produced in its breeding programme.
- IRRI will not seek intellectual property protection on the breeding lines, elite germplasm and parental lines emanating from its breeding programme.

- IRRI recognizes that the private sector is likely to play an important role in the development of hybrid rice technology.
- IRRI will provide parental lines (and other elite materials) to both the public sector institutions and the private organizations on the understanding that:
  - The material is not intended for exclusive use by any single organization.
  - IRRI retains the right to distribute the same material to other organizations.
  - The use of IRRI materials will be recognized when a hybrid rice variety is released.
  - IRRI's collaboration with profit-making organizations for the development of hybrid rice technology will proceed after consultation, with the authorities.

In this context, it is felt that conceivably more strenuous efforts on hybrid research, especially by IRRI and CIMMYT on rice and maize, respectively, would go a long way in promoting hybrid technology. ICRISAT has already placed a strong research agenda for sorghum and pearlmillet hybrid programmes. These centres need to enhance the capabilities by building the National Agricultural Research System (NARS) with more aggressive efforts.

Environmental Implications of Hybrid Technology

The future of world agriculture belongs to hybrids. In fact, homogeneous population of heterozygous plants will replace cultivation of homogeneous population of homozygous plants (many self-pollinated crops) and heterogeneous populations (most cross-pollinated crops) in the next 25 years. The trend is unambiguous, therefore for faster progress, India should evolve a pro-hybrid policy in its crop improvement programmes as we need hybrids more than most other countries. The country is fortunate enough to have the scientific manpower to take leadership in organizing research programmes on this direction. Dr. C. T. Patel of Gujarat Agricultural University blazed a new trail, in 1970, by releasing the first ever commercial hybrid of the prized cotton species. Today, nearly 2.4 m ha of high quality cotton area planted with hybrid seeds is making the country world's second largest producer of cotton fibre.

Similarly, Barber of Sugarcane Breeding Station, Coimbatore developed the hybrid of sugarcane CO 205 in 1918. A different kind of hybrid varieties helped India to become the second largest producer of sugar in the world.

The concept in both the cases is to synthesize favourable gene combinations into highly heterozygous varieties, and then maintain the genotype intact, through repeated production of F_1 seeds in one case and asexual reproduction in another.

The Indian rice improvement programme has responded well, and the country can look forward to an array of rice hybrid varieties in the next decade with at least 25 per cent of rice area under hybrid varieties by 2010.

It is recognized that the most important need of the day is to create high genetic potential in management responsive yields and improve the agronomic management in farmers fields with the application of modern farm inputs like fertilizer and irrigation. The plant breeders responded to this new challenge by making use of plant type genes to alter the traditional cultivars. The dwarf high yielding varieties of wheat and rice have been the product of a strategy of this kind.

The hybrid varieties can help to take the modernization process a step further. The hybrid varieties perhaps offer their greatest potential at present for the transformation of Indian horticulture as well.

In the event the world agriculture moves increasingly towards hybrids, what will be the environmental implications? An obvious one will be the increasing genetic uniformity of different crops as F_1 is extremely uniform showing a high degree of heterozygosity with extreme homogeneity. Coverage of large areas with limited number of hybrids will make the crop vulnerable to the build up of disease and pest epidemics. Therefore, it is predictable that the hybrids may impose same problem as faced by adoption of high yielding varieties of wheat and rice.

The other major environmental implication of the hybrid technology is that, in general, hybrids express their full yield potential under intensive agronomic management based on the use of large doses of modern farm inputs like chemical fertilizers. Hybrids have little place in low input agriculture *i.e.* traditional system of farming. Population explosion in many developing countries left no alternative to increase in per unit production by growing management-responsive varieties. This would have some serious environmental implications in terms of pollution of land, water and soils, with maximum exploitation of resources to sustain today's vastly increased populations. The answer lies in developing a more modern agricultural technology based on recent advances in molecular biology.

Recent advances in science offer the possibility of making agriculture highly productive, efficient in terms of use of resources and environmentally safe, but this requires evolution of new research priorities and restructuring of the Indian national agricultural research system.

The productivity levels of rice, wheat and maize have increased, but achievement in sorghum and pearl millets is not up to the mark. On the contrary, to realize the genetic potential yield of these crops, especially through the use of hybrid technology, needs adoption of suitable strategies.

Among various technological options available for the improvement of productivity, perhaps heterosis-breeding approach is the most successful one, since this technology has already been exploited extensively in several crop plants. In India, distinct advantage of hybrids of maize has been experienced on a large scale with considerable productivity gains. Recently, hybrid wheat and sesame has been released for commercial production in China.

From these developments, one gets convinced that hybrid technology could successfully be employed for further improvement of productivity of important food crops. However, considerable gap exists between the national productivity levels and the realizable as well as genetic yield potential of hybrids.

Strategy for the Future

Suitable strategy on hybrids is necessary to overcome existing constraints by improving production and productivity of major cereals. Following action would be helpful for improvement in the present situation:

- National policy concerning adoption of hybrid technology is required to determine the extent of impact.
- Greater thrust on research and development efforts for development and adoption of hybrids at field level, as present investments are inadequate.
- Organized seed production system of hybrids for higher replacement rates to improve the production.
- Besides the public system, private seed sector has to be encouraged to play an effective role for appropriate and efficient delivery system of hybrid seed.
- The pace for adoption of hybrid technology should be accelerated.
- Presently regular flow of information on hybrids is lacking, therefore a strong collaboration for seed production and promotion of hybrid technology based on experiences and successes gained is mandatory.
- Efficient and systematic testing procedure of hybrids should be established.
- Exchange of germplasm, information on hybrid technology, seed production and marketing between countries should be encouraged.
- Human resource development for economical seed production and supply of hybrid seeds at competitive prices.

- Cooperation among public and private seed organizations, especially to draw strength from each other in view of comparative advantages.
- Advice concerning policy and strategy for promoting R&D efforts in hybrid technology and development of seed sector.

Hybrid Crops: Plans & Policies

In the late fifties, high yielding hybrids of flint maize were released for mass scale cultivation, which initiated the use of hybrids in the country. There has been a radical transformation in the agriculture scenario in the country in the sixties, and hybrids in many crops have played an important role in increasing productivity as the hybrids of sorghum, pearlmillet and cotton showed inherent superiority over the popular local varieties.

However, the cultivation of sorghum and pearlmillet did not much impact on improvement in average yields over the country as a whole because these crops are primarily grown under rainfed conditions with considerable variation in rainfall pattern, both spatially and in time.

With the spread of hybrids and their acceptance, the demand, especially of sorghum, pearl millet and cotton had steadily increased in the central and western parts of country. Subsequently, the Govt. of India established State Seeds Corporations to enable production of sufficient quantities of hybrid seed to meet the requirements on a systematic basis, as hybrid seeds have to be replaced each time the commercial crop is raised. Even then, increasing demand of hybrid seeds often encountered acute shortages partly due to mismatch between the availability of foundation seeds and demands of desired hybrids for which seed producers wanted to take up seed production programme.

Emerging Issues

Seed of several crops have reached almost every nook and corner of the country. In this process, bulk of seed production takes place under contracts between the seed companies and individual seed growers mainly due to small size of holdings and ceiling on land holding. That is an extremely important characteristic of the Indian seed industry.

The introduction of hybrids meant a radical and qualitative change in the seed production system. Technologies associated with hybrid seed production, such as staggering for nicking, planting ratio, supplementary pollination, isolation, rouging, pollen shedder, maintaining disease free crop, harvesting etc. have to be calibrated according to crop and place of seed production. Seed processing technologies have also been continuously modernized and upgraded over a time.

Present state of seed development programme raises the following important policy issues:

- The role of the public and private sector research in evolution of new hybrids of different crops.
- Implementation of new policy on seed development announced in 1988 and changes required as per its objectives
- Performance assessment of the hybrids in different agro-climatic conditions along with their popularization through the extension system.
- Production, quality control, distribution and pricing of seed.

Almost all over the world, in the initial stages, the role of the public sector research institutions or public-funded research programmes played a major role in generating technologies for commercial application. However in the recent decades, the part played by the private industry in generation of technology for production and products has been very significant and is increasing. The country has established a very large research network working under the aegis of the Indian council of Agricultural Research (ICAR) and the State Agricultural Universities. The policy contemplated that the parental material should also become available for multiplication within the country leading to production of hybrids on an adequate scale to cater to the demands.

Hybrid Seed Production and Popularization: Issues and Strategies

A choice between an inbred variety or hybrid for any particular situation is governed by the properties of the material developed for the particular situation in conjunction with availability of its seed at the right time. General overview of the advances in different crops indicates the higher yield potential of hybrids. For example, the plateau of yield levels in non-hybrid wheat and rice (although rice hybrids have been developed in India that may significantly increase production) on one hand and the tremendous potential witnessed in new hybrids in maize, pearlmillet, sorghum, sunflower, castor, cotton etc.

The success of hybrids, *viz.*, pearlmillet in arid region of western Rajasthan; sorghum in Maharashtra, Andhara Pradesh, Madhya Pradesh and Tamil Nadu; cotton in Gujarat, Maharashtra and South India; castor in Gujarat and maize and sunflower all over the country bestowed the required confidence to farmers for opting hybrids against pure line varieties.

It is necessary to build up the confidence of the farmers in hybrid technology, by way of continuous efforts on hybrid research in promising crops in conjunction with well-planned efforts for hybrid seed production and popularization. Looking in to the worldwide hybrid research, the

22nd century is expected to witness an all round acceptability of hybrids in a large number of agriculturally important crops. It is realized that with conscious breeding efforts hybrid technology can profitably be exploited in stress environments as well, and at times it will change the grey rainfed areas to green belt.

Compiled information on certified hybrid seed production in Andhra Pradesh, a major seed producing state, clearly reveals the excessive demand of hybrid seeds of sorghum, maize, pearlmillet and cotton. The state has yet to achieve targets in case of hybrid castor and sunflower.

In India the total production of breeder seed of parental lines of hybrids far exceeds the indent. It indicates the initiative taken for popularization of promising hybrids even without firm indents. A perusal of breeder seed indents reveals small share of newly notified materials. It is also noticed that even 2–3 years old hybrids are indented on a very small scale from some of the states. Thus, the awareness of newly developed materials is only up to small extent that requires an aggressive approach for popularization. On the other hand, the private sector mainly markets the hybrid seeds, as fresh seed is required every year. And its demand is gradually escalating with the declining trends in public bred hybrids.

Seed Requirements for Meeting Production Targets

Availability of quality seed of the required hybrid at the right time would remain a critical factor in achieving the targets. Therefore it is imperative to focus on the following general issues related to seed production and supply along with the desired strategies for overcoming difficulties before addressing the issues specific to hybrid seed production and popularization

1. Need for significant increase in seed production and supply.
2. Establishing a database for seed production and supply.
3. Advance planning and working on a rolling plan concept.
4. Involving more agencies in seed production system by encouraging potential organizations on a localized basis, like seed cooperatives and voluntary organizations in M.P. It can greatly augment certified/ quality seed production activity in the country. Identification of promising village *panchayats* for this activity is worth examining in the light of experience gained from the *seed village concept* implemented in the initial period of organized seed production in the country.

Alternative Areas for Seed Production

A sizeable portion of seed production activity is confined to the southern states *i.e.* Andhra Pradesh, Karnataka and Maharashtra. This is certainly not a desirable situation in a long term perspective planning. Outbreak of any diseases or pests in these areas can give a severe jolt to production activity of this critical resource. The examples of extensive damage by American bollworm and white fly pests of cotton in South India and downy mildew in sunflower in Maharashtra are still intact in memory.

Dry climatic regions of Rajasthan with irrigation potential would be far more rewarding from disease management point of view, hence worth pursuing. This issue needs to be specifically examined by a group of experienced persons involved in seed research and production along with enterprising people from the private seed sector. A comprehensive seed atlas has to be developed for the country with potential areas for seed production in off-season.

Contingency Planning and Buffer Stocking

Contingent planning and buffer stocking should become an essential element of planning and execution of seed programmes must enobige.

- Seeds of hybrids with wide adaptability and/or ability to fit into different cropping systems for utilization in unforeseen situations.
- Early maturing hybrids should figure in contingent planning to cater the emergent area-specific situations like drought, flood, etc.
- Seeds of hybrids/inbred varieties with less demand may be produced for the next few years and buffer stocked as a well thought of strategy.

Popularisation of Public Sector Materials

It is unfortunate that very little time and effort had been devoted for popularization of hybrids released from the public sector. Certainly, it is an undesirable situation as public sector hybrids have evolved through reasonably well-planned efforts of institutions and research stations spread all over the country and released through a worldwide-acknowledged system of multi-location coordinated testing.

Hybrids/inbred varieties released from the public system still occupy the dominant share in India's agricultural production but it takes an awfully long period of 7–10 years before the newly released materials become available to farmers. The larger gap between average crop productivity at the state or national level and realizable potential can be reduced only through efforts in popularization of newly developed hybrids. In this endeavor, the Govt. of India may develop a system where seed of newly released varieties in the very first year of release could be distributed in small packets up to village *panchayat* level to bring much needed awareness in short time.

Specific Attention for Neglected Crops/Commodities and Regions/ Situations

A number of good hybrids have been released by the public system that requires special attention by the Government for popularization. Likewise, for popularization of floriculture and fruit trees limited efforts have been made by the Government of India. The major issue is whether the Government should largely leave these commodities of great economic importance for business by the private sector only? Many hybrids in different vegetable and fruit crops developed in the public sector require special popularization efforts.

Requirement of Skilled Manpower

Planning for skilled manpower and developing a system for refresher courses at regular intervals is a critical issue to be tackled. In case of hybrid seed production, the skill requirement is obviously much higher. Planning and system for meeting the training requirements should, therefore, be immediately focused. Simultaneously, a study for examining the educational qualifications and experience of involved personnel is desirable. Ad-hocism in this critical area rather than having a long term perspective planning has been an important reason for non-achievement of targeted production level of certified/quality seed.

Advent of Hybrid Culture

Rice, mustard and maize hybrids from the private sector are reportedly on the anvil. A worldwide explosion of private sector efforts in agricultural research would mostly bring hybrids in both conventionally hybrid and non-hybrid crops. In India, the upcoming private sector coupled with increased efforts in large public sector research and development of infrastructure may bring miracles, but further improvements in the system of governmental support and encouragement with farmers' interest uppermost in any policy initiatives are necessary. A vertical expansion in production through encouragement of hybrid research and development would thus be possible provided the government's overall seed research, production and supply management systems are appropriately tailored for the required modifications to give signal to all concerned.

It is, therefore, essential that issues and strategies pertaining specifically to hybrid seed production and popularization be highlighted separately.

Vision for Future

The responsibility to develop hybrids for different ecosystems and less profitable crops is owned by public sector, as the private sector is not ready to spend in research with high gestation period and low/unassured immediate returns. Vision is required to share responsibilities between private and public sectors, as in future public sector alone may not be able to face this challenge. At the same time, the relevance of public sector and up and coming private sector requires continued working in crops/situations of interest of private sector as well. Desirable results require joint efforts.

Policy Support for Hybrid Seed Development

At national level, public sector seed corporations have not made any significant impact in the field of hybrid seed production. This being a skill requiring procedure compared to inbred varieties, requires sensitization of these corporations, through planned efforts in human resource development besides policy initiatives encouraging hybrid seed production and popularization. Healthily competition between private and public sector could be rewarding. Issues of subsidy and incentive pricing for hybrid seeds are worth examining in view of anticipated benefits.

Considering Intellectual Property Regimes in Agriculture

Even with in-built security in hybrids, the apprehensions about theft/misuse of parental lines have been expressed by the private sector. On the other hand, parental lines of hybrids developed in the public sector are available freely to both private and public sector agencies without discrimination for the benefit of the farmer. Thus, in the first instance, a view has to be finalized on sharing of parental lines of public hybrids, and simultaneously a system for registration of parental lines is necessary to specifically acknowledge and reward the concerned company or institution. The on-going debate in the country on Intellectual Property Regimes (IPR) in agriculture has revealed that a system of Plant Breeders' Rights with exemption for research and farmers is based on national interest.

Research as an Integral Part of Private Seed Sector Activity

The idea of healthy competition between private and public sector efforts would be possible with equal supportive role of private sector on varietal development. In the present situation, only a few private seed companies have sound research base, whereas most are benefiting from efforts of public sector without sharing benefits. Initiatives for research in the private sector would complement/supplement/augment efforts in public sector to meet India's large requirements.

Gearing up Seed Technology Research

Seed technology research is still in infancy in India. Ideally, any research product should release a complete technology package, with of details meeting the expected specific/diverse requirements. For example, release of hybrids without package for agronomy of parental lines and economics of seed production would bring serious obstacles in popularization efforts. Preparation of technology capsules is an enormous task as it varies depending on regions, seasons and situations and thus should not be left as the sole responsibility of public sector research system alone. Seed producing agencies should come forward and play equal role in seed technology research in important areas like agronomy for hybrid seed production, quick genetic purity tests including biochemical/molecular approaches, seed treatment, storage, processing, packaging, etc.

Effective Inter-Project Linkages

Inter-project linkages between All India Coordinated Seed Technology Research Project and a few All India Coordinated Crop Improvement Projects, *viz.*, maize, sorghum, pearlmillet, rice, pigeonpea, sunflower etc. have been established to jointly solve crop-specific problems of seed production, storage, processing, etc. The effort needs further strengthening. This is a critical grey area coming in the way of popularization efforts for newly released hybrids. Besides crop-specific issues, which would vary from crop to crop, issues like characterization of released materials to meet international requirements, effective protection of released materials through development of identification keys in the form of appropriate database, and maintenance breeding need immediate streamlining, with fixation of specific responsibility for each task.

Reviewing Seed Standards

Seed standards are prerequisite for seed quality control. Seed certification standards enforced in India are largely drawn from information available from other countries with subsequent moderation based on practical difficulties posed by local seed producing agencies. Among various seed standards, seed-borne diseases assume specific importance, being the most important reason for rejection of seed lots. The high standard of genetic purity (95%) for hybrids discourages seed producers.

Improving Seed Quality Enforcement

An effective system for monitoring seed quality, right from breeder seed up to certified seed, has to be in place as a support for seed quality enforcement. Important issues like educational qualification, experience and regular training of officials involved in seed certification, testing and quality law enforcement, up gradation of seed testing laboratories in states, and registration of seed dealers, merchants, packers and licensed crop inspectors need to be critically reviewed, followed by appropriate strategy for improvement. Grow-out test for ascertaining genetic purity, particularly of hand-pollinated hybrids, is an essential requirement. It is necessary to work out modalities of this task for implementation in all hybrids in an infallible manner.

2 Strategies for Commercialization of Public - Bred Hybrids of Rice in India

M.P. Pandey and S.R. Dhua*

ABSTRACT

Significant cross pollination in the cytoplasmic male sterile lines of naturally self-pollinated crop of rice has made exploitation of heterosis possible to boost the national productivity level. Prospects of exploitation of inter sub-specific heterosis using thermo-sensitive genic male sterility with genes for wider compatibility has been explored with most encouraging results from two-line hybrids developed and tested in multilocation trials under All India Coordinated Rice Improvement Project. However, apomixis is the ultimate genetic tool to develop true breeding hybrids and facilitate commercial exploitation of heterosis by even resource poor farmers of the developing world. Similarly impact of vegetative propagation of rice hybrids is expected to felt during the next few years. Public-private partnerships in large scale hybrid seed production has to be given a boost for universal adoption of the technology.

Key words: Commercialization, grain quality, hybrid rice, linkages, management, marketing, public bred, quality control, regulatory mechanism, seed production, stresses, vegetative propagation

National and International Crop Scenario

Rice occupies a pivotal place in the global food and livelihood security systems. Of the annual world production of 596.485 m tons (paddy) from 155.128 m ha in 1998–'99, Asia produced 540.621 m tons (90.63%) from 138.563 m ha (89.32%). Average rice yield in Asia was 3.9 T ha^{-1}, compared with 3.8 T ha^{-1} in the world, 6.4 T ha^{-1} in Japan, 6.3 T ha^{-1} in China, 2.9 T ha^{-1} in India, and 10.1 T ha^{-1} in Australia. Per capita consumption of rice in Asia ranges from 132 to 449 g per day. The world population will reach

* Crop Improvement Division, Central Rice Research Institute (CRRI), Cuttack - 753 006.

seven billion near the year 2010, eight billion near 2020, where 82% will live in the developing countries. The UN forecasts that the world population will reach 9.4 billion by 2050. It is projected that there will be only 1.2 billion more people in the developed regions of the world, compared to 8.2 billion in the currently less developed countries, with 5.5 billion of them in Asia and 2 billion in Africa. The world must develop the capacity to feed 10 billion within the next 40 to 50 years, predominantly within Asia and Africa.

As regards India, rice is a staple food, providing 43% of calorie requirements for more than 70% of the population. Among all the crops grown in India, rice occupies the largest area (about 44.6 m hectares) which is also the largest in the world among all the rice growing countries. The country had a production of 139.14 million tons of paddy in 2006–07 with a productivity of 3.1 T ha^{-1}. The area under rice increased by 1.5 times, from 30 m hectares in 1950 to about 44 million hectares now; while during the same period the production increased by more than four times, *i.e.* from 22 m tons to 92 m tons of rice (cleaned). The productivity thus increased three folds, *i.e.* from 0.7 T ha^{-1} to 2.1 T ha^{-1}. This spectacular progress helped India to achieve self-sufficiency in rice production during the early 1980's.

Over the past four decades, technological innovations in rice breeding became a frontier for increasing rice productivity in the rice-growing world. The development and massive promotion of the so-called "modern" or "high-yielding varieties" with combined package of technologies such as chemical fertilizers, pesticides, mechanized land preparation and irrigation transformed the landscape in most rice farming communities. Rice yields initially increased rapidly. However, the yield of modern varieties reached a plateau in the 1980's and required further breakthrough in its yield potential.

Scope and Importance of Hybrid Rice

Increase in rice productivity is highly essential to keep pace with the increase in population and decrease in the available land for rice cultivation in the country due to conversion of paddy lands into non-paddy cultivation for agricultural diversification and also the conversion of agricultural land to non-agricultural purpose. The increase in rice production can be achieved either by improving the existing HYVs or developing heterotic hybrids.

It has been observed that the hybrids can give 15–30% yield advantage over the best commercial varieties. The scientists have shown hybrid rice to be the globally and also in the country the best technological option readily available to address the dwindling rice production because of the following facts:

- Yield levels of modern varieties developed during the Green Revolution era have reached a plateau.

- Increasing demand for rice especially in developing countries.
- Better performance of hybrid rice under adverse conditions like drought and saline conditions.

China remains to be the country with the widest scale of hybrid rice cultivation across Asia covering about 15 million hectares. Outside China, hybrid rice cultivation have been increasing in Vietnam, Bangladesh and the Philippines during the recent years (Table 2.1).

Table 2.1. Areas of hybrid rice cultivation in Asian countries

Country	Area of adoption (hectares)
China	15,215,000
India	900,000
Bangladesh	40,000
Vietnam	480,000
Philippines	208,342
Indonesia	20,000
Myanmar	50,000

The Government has heavily subsidized promotion of hybrid rice in China. Farmers were provided with free seeds, supported by subsidies on fertilizers and pesticides, and were instructed to plant hybrid rice. There was a ready market for hybrid rice as hybrid rice seed and grain were procured by the state. There was such massive adoption of hybrid rice in China that in 1992, more than 50% of its irrigated rice area was planted with hybrid rice.

Inspired by the success of China, IRRI in collaboration with the Peoples' Republic of China, Asia Pacific Seed Association (APSA), FAO and NARS of several Asian countries initiated a project to promote hybrid rice technology outside of China. These efforts led to the commercial release of hybrid rice varieties in the early 1990's and were subsequently promoted as national hybrid rice programs.

These efforts received various financial and technical support projects from both FAO and ADB. IRRI became the main source of parental lines of hybrid rice to NARS and even to the private sector. APSA facilitated the entry of private sector in hybrid rice dissemination and complementation of private seed interests with national hybrid rice programs.

Achievements in the Development of Hybrids

While modern varieties were spreading in the 1970's, scientists in China had successfully developed the hybrid rice technology capable of out-yielding the modern varieties. With subsequent massive promotion of the technology, hybrid rice enabled the country to become rice self-sufficient and able to feed it's more than a billion populations.

Encouraged by the success of the technology in the Peoples Republic of China in 1974, the Indian Council of Agricultural Research in December 1989 initiated a time bound, result oriented project in a network fashion on the "Development and Use of Hybrid rice Technology" in a mission mode. With the concerted research work, the country developed four rice hybrids in 1994 (*viz.*, APHR 1, APHR 2, MGR 1 and KRH 1) from public sector within a short span of five years Subsequently, hybrids CNRH 3, DRRH 1 and KRH 2 were also released. A total of 29 rice hybrids have been released till 2007. The salient features of these hybrids are given inTable 2.2.

Table 2.2. Rice Hybrids released in India

Variety	Year of release	Duration (days)	Yield (t/ha)		Superiority over check (%)	States for which released
			Hybrid	Check		
1. APHR 1	1994	130–135	7.14	5.27	35.4	AP
2. APHR 2	1994	120–125	7.52	5.21	44.2	AP
3. MGR 1	1994	110–115	6.08	5.23	16.2	TN
4. KRH 1	1994	120–125	6.02	4.58	31.4	KA
5. CNRH 3	1995	125–130	7.49	5.45	37.4	WB
6. DRRH 1	1996	125–130	7.30	5.50	32.7	AP
7. KRH 2	1996	130–135	7.40	6.10	21.3	AP, KA, TN, Tripura, MH, Har, UT, Raj
8. Pant Sankar Dhan 1	1997	115–120	6.80	6.20	9.7	UP, UT
9. PHB 71*	1997	120–125	6.25	5.20	20.2	TN
10. CORH 2	1998	115–120	7.10	4.90	44.9	TN
11. ADTRH 1	1998	125–130	6.64	4.89	35.8	MH
12. Sahyadri	1998	125–130	6.15	4.94	24.5	UP
13. Narendra Sankar Dhan 2	1998	130–135	7.86	6.14	28.0	Har, UP, TN
14. PA 6201*	2000	125–130	6.18	5.03	22.9	Eastern States, AP, KA, TN
15. PA 6444*	2001	135–140	6.11	4.91	24.4	UP, Bihar, Tripura, Orissa, AP, KA, MH
16. Pusa RH 10	2001	120–125	4.35	3.11	39.9	Har, Punjab, Delhi, WUP
17. RH 204*	2003	120–126	6.89	5.62	22.6	AP, KA, TN, Har, UT, Raj
18. Suruchi 5401*	2004	130–135	5.94	4.97	19.5	Har, AP, KA, MA
19. Pant Sankar Dhan 3	2004	125–130	6.12	4.99	22.6	UT
20. Narendra Usar Sankar Dhan 3	2004	130–135	5.15	3.86	33.4	Saline alkaline areas of UP
21. DRRH 2	2005	112–116	5.35	4.85	24.9	Har, UT, WB, TN
22. Rajalaxmi (CRHR 5)	2005	130–135	5.71	4.47	27.9	Orissa

Table. 2.2. *Contd.*

Variety	Year of release	Duration (days)	Yield (t/ha)		Superiority over check (%)	States for which released
			Hybrid	Check		
23. Ajay (CRHR 7)	2005	130–135	6.07	4.47	35.9	Orissa
24. Sahyadri 2	2006	115–118	6.50	5.20	25.0	MH
25. Sahyadri 3	2006	123–126	7.50	6.40	17.0	MH
26. HKRH 1	2006	135–139	9.41	8.17	15.2	Haryana
27. CORH 3	2006	130–135	6.15	4.90	25.5	TN
28. Indira Sona	2006	125–128	6.80	5.45	24.7	Chhattisgarh
29. JKRH 2000	2007	130–135	6.90	5.70	21.0	WB, Orissa, Bihar

**Developed by private sector.*

Six of the 29 hybrids were developed and released by the private sector organizations with strong R&D wings. Also, 21 of these hybrids have been released by different State Variety Release Committees, while eight hybrids (PHB 71, PA 6201, PA 6444, RH 204, Pusa RH 10, KRH 2, Suruchi 5401 and DRRH 2) are released by the Central Variety Release Committee. Many of the hybrids have shown considerable yield advantage over the comparable best checks available for different regions. Most of the hybrids are in the mid-early to medium duration groups while only hybrids Pant Sankar Dhan 1 and Sahyadri 2 are in the early group. One of the hybrids, Pusa RH 10 is an aromatic hybrid with superfine grain while hybrids, Ajay and Rajlaxmi are suitable for irrigated as well as shallow rainfed lowlands. Hybrids suitable for the *boro* season and aerobic situations have also been identified in the AICRP program. Some of the recently released hybrids have good grain and cooking quality characteristics.

Grain Quality Considerations

Grain quality characteristics (physical, milling and olfactory) are important considerations for determining consumer acceptance of any hybrid. The quality characteristics of the released hybrids (Table 2.3) revealed that the grain quality of the second phase hybrids is comparable or even superior to most of the popular varieties such as IR 64, Jaya, Sarjoo 52 etc. However, only few hybrids are comparable to the premier quality varieties like Sona Mahsuri, White Ponni, BPT 5204 etc. Lower head rice recovery, stickiness and mild aroma (for southern states only) are the negative characters affecting the grain quality of the hybrid. The present problems of grain quality of these hybrids are primarily due to narrow genetic base of the CMS lines being used. Efforts are underway to develop well-adopted indigenous non-aromatic CMS lines, with medium slender grains having higher head rice recovery and intermediate amylose content. By choosing appropriate parental lines, it would be possible to develop hybrids of desirable grain quality in near future.

Table 2.3. Grain quality characteristics of the released hybrids and check varieties

Hybrid or variety	Grain type	Milling (%)	Amylose content (%)	Alkali value	Tenderness on touching (after cooking) (1–5)	Stickness (after cooking) (1–5)	Taste (after (cooking) (1–5)
DRRH 1	LS	65.2	24.0	4.2;4.3	4.4	4.3	3.4
CNRH 3	LB	69.3	27.8	3.1;3.8	3.0	4.3	2.7
MGR 1	MS	67.6	31.4	–	3.4	3.7	2.6
CORH 2	MB	60.5	26.3	3.6;4.5	3.6	3.3	2.8
ADTRH 1	LS	70.2	24.7	3.9;4.6	3.7	4.3	2.8
APHR 2	LS	69.0	27.5	2.3;2.6	4.0	2.8	2.6
KRH 2	LB	72.6	27.5	2.0;2.0	3.7	4.0	2.9
Sahyadri	LS	67.6	26.2	–	4.3	3.2	3.1
PA 6201	LS	69.7	24.0	2.4;3.3	3.4	2.7	2.8
PHB 71	LS	66.0	24.2	2.4;3.3	3.4	4.3	2.8
Jaya	SB	71.7	27.5	7.0;7.0	3.8	3.4	2.8
Rasi	SB	68.5	30.8	7.0;7.0	3.8	4.5	2.6
Swarnadhan	SB	60.6	29.1	7.0;7.0	4.0	4.6	3.2
Samba Mahsuri	MS	69.0	23.8	3.0;4.0	4.1	4.5	3.6
Vikas	LS	66.5	27.2	6.0;6.0	2.6	1.1	2.0

2: Mod hard, 3: Mod soft, 4: Soft

2: Tasteless, 3: Desirable, 4: Good

LS—Long Slender; LB—Long Bold; MS—Medium Slender; MB—Medium Bold; SB—Short Bold

Resistance to Major Biotic/Abiotic Stresses

For stable performance of the hybrids across locations/seasons, it is necessary that the hybrids possess resistance/tolerance to some of the major pests and diseases. Hence hybrids under development are regularly screened against major pests such as stem borer, BPH, WBPH and gall midge and diseases like bacterial leaf blight, blast and sheath rot. Resistance/tolerance of hybrids to some of the major pests and diseases is given in the Table 2.4.

Hybrids released in the beginning were not resistant/ tolerant to any of the pests and diseases. However, the hybrids released recently show either resistance or tolerance to some of these pests and diseases. For example, hybrid HRI 120 (6444) released in 2001 has resistant to blast and WBPH and tolerant to BLB, sheath rot, BPH and gall midge. Now increasing emphasis is being given to incorporate resistance to major pests and diseases in the parental lines and hybrids. Some of the hybrids like PHB 71, CORH 2, Narendra Usar Sankar Dhan 3 etc. perform well under abiotic stresses like saline/alkaline as well as under reduced moisture conditions.

Table 2.4. Resistance/tolerance of hybrids to major pests and diseases

Hybrid	Resistance	Tolerance against
KRH 1	–	Blast
DRRH 1	Blast	–
KRH 2	Blast, Sheath rot	–
Sahyadri	–	BLB
Narendra Sankar Dhan 2	Blast	BLB, Sheath rot
PHB 71	–	Blast, BLB, BPH
PA 6201	Blast	BPH
HRI 120 (6444)	Blast, WBPH	BLB, Sheath rot, BPH, Gall Midge
Pusa RH 10	–	BLB, BPH
RH 204	–	BPH, WBPH
Ajay	–	BLB, Blast, SB, GM, BPH, WBPH
Rajlaxmi	–	BLB, Blast, SB, GM, BPH, WBPH

Issues/Programs in Hybrid Rice Research and Management Strategies

Hybrid rice in India is ready for a major take off in the implementation of the National Food Security Mission programme of the Government of India during XI Plan. However, the issues related to large scale seed production using the present three-line system would remain a constant need till we are able to produce effective two-line hybrids using environmentally sensitive genetic male sterility and/or use of apomixis technology of commercial exploitation of one-line hybrids.

The following issues remain important till they are resolved with greater accuracy:

1. Hybrid seed production remains risky, technical and expensive because of the following problems:
    - Synchronized flowering of the parental lines
    - Higher production input (GA_3 and labour) cost
    - Parental susceptibility against (biotic/abiotic) stresses.
2. Potential of hybrid seed yield not well established/documented under farmers conditions:
    - Ideal location/season not identified
    - Inadequate training in seed production technology
    - Non-availability of hybrid seed in adequate quantity.
3. Inadequate demand of hybrid seed
    - Real yield potential of hybrids not extensively demonstrated

- Inconsistence performance of the hybrids due to improper crop management
- Poor grain quality, *e.g.*, presence of aroma in IR 58025 A based hybrids is not liked in the southern region. Low milling recovery of some hybrids is responsible for lower price paid by the millers to cultivators
- Higher seed cost is deterrent as CMS and restorer parents have lower productivity and poor combining ability and out-crossing potential of the parental lines is the key factor for low seed yield for commercial grain crop.

4. Genetic purity of hybrid seeds
   - Non-availability of pure nucleus seed stock of parental lines on sustainable basis
   - Lack of purity of foundation seed stocks of parental lines due to poor isolation and inadequate monitoring of genetic purity of breeder seed. Creation of consortium for the multiplication of foundation seed stocks of parental lines of public bred hybrids could provide quality foundation stock in adequate quantity to the seed growers (public/private)
   - Lack of rolling plan for the multiplication of public bred hybrids either by the Department of Agriculture and Cooperation (Ministry of Agriculture, Govt. of India) at the Central level or the Department of Agriculture at the level of States
   - Very poor integration of research, seed production and technology transfer departments of the government
5. Inadequate sensitization of administrator/policy makers about the hybrid rice technology.

Management Strategy for Large Scale Seed Production

Management strategies related primarily to research, production and policy issues as detailed below:

Research Management strategies

1. Greater thrust on parental line improvement and methodology development:
   - Improvement of female parents for higher general combining ability and the pollen parent for specific combining ability needed
   - Improving productivity of parents and their resistance against major diseases and pests

- Enhancing higher out-crossing potential of parental lines
- Development of system of two line hybrid breeding with the prospects of exploitation of inter sub-specific heterosis.

2. Further refinement of agri-management techniques for super high yield of parental lines and hybrids in production plot.

   Adopt cultural practices for obtaining 4.0–4.5 million effective panicles/ha of female parent. The practices recommended in China are:

   - Raise vigorous and multi-tillered seedlings (15 g/m^2)
   - Supply of sufficient basal fertilizers in the nursery
   - Early and heavy top dressing
   - Timely transplanting at 5th leaf stage
   - Appropriate row ratio
   - Appropriate field management.

   Study on the development of agro-technology for hybrid seed production with Pant Sankar Dhan 1 at Pantnagar for two years (Pal and Pandey, 2003) revealed the highest average yield of 3.9 t ha^{-1} for the hybrid and 7.5 t ha^{-1} for the pollen parent. The best agro-technology treatment included lower seed rate (25 g/m^2), deep seeding (5.0 cm) and three split applications of N (60% basal + 30% top dressing at 15 DAS + 10% at 25 DAS) in the nursery and four splits of N (30% basal + 20% top dressing at 10 DAT + 30% at 10 days before booting) and transplanting two seedlings/hill produced on an average 376 panicles/m^2 (3.76 million panicles/ha), 291.3 million spikelets/ha, 24.6 cm long panicles with 1000-seed weight of 22.25 g and better panicle exertion (77.42%). These findings have great implications in increasing hybrid rice seed yield in the country.

3. Develop techniques for higher out-crossing rate:

   - Seed production under ideal location/season, sufficient sunshine and mean temperature of 26°C with the difference between day/night temperature be preferably 5–7°C, RH 70–80% and moderate wind velocity
   - Use fertilizers to promote higher spikelet formation and pollen load and density
   - Apply GA_3 skillfully at higher dose and concentration
   - Practice light flag leaf cutting with higher dose of GA_3
   - Regulate anthesis time for proper synchronization of both the parents.

4. Determining the proper isolation distance and synchronization of parental lines across locations in the country
5. Finding cheaper substitute(s) of GA_3 either fully or partially to economize the cost of seed production. Studies have shown the feasibility of substitution of GA_3 up to the extant of 15–30 g/ha without yield loss with a cheaper chemical composition (CC) comprising urea 0.1% + boric acid 0.1% + zinc sulphate 0.1% and potassium biphosphate (Table 2.5).

Table 2.5. Effect of GA_3 substitution by cheaper chemical composition (CC) on hybrid seed production of Pant Sankar Dhan 1

Treatment	Yield (kg/ha)		Average
	1st year	2nd year	
Control (no GA_3)	1104	1193	1148
GA_3 45 g	1817	1867	1842
GA_3 30 g + CC	1755	1772	1763
GA_3 15 g + CC	1643	1731	1687
Overall mean	**1580**	**1689**	**1634**

During the third year, application of GA_3 (30g) + NAA (50g) yielded 3164 kg/ha and was at par with highest yielding treatment of 90g GA_3 with yield of 3201 kg/ha.

Strategies for Commercialization

The availability of quality seed at a reasonable price is a crucial factor behind the large scale adoption of hybrid technology for any crop. This has been demonstrated by the success of hybrid varieties of maize, pearl millet, sorghum, sunflower and cotton, all of which are supported by economical and efficient seed production and distribution systems in the countries where the hybrids are commercialized. Chinese success in the large scale adoption of hybrid rice can also be attributed to the efficient and economical hybrid seed production and distribution system, organized by the state-owned seed industry.

Hybrid rice seed production technology is considered labor- and knowledge-intensive. It involves various risks, especially in the early stages when seed producers are still lacking in experience. Typical problems are poor synchronization of the parental lines, and unfavorable weather. At present, commercial production of hybrid rice seed outside China is limited to India, Vietnam, the United States and the Philippines. Seed growers will engage in hybrid rice seed production only if it is more profitable than inbred varieties, and if the additional profit is enough to compensate for the greater risks and level of skill involved.

Studies conducted in the Indian states of Andhra Pradesh, Karnataka and West Bengal showed that on average, labor accounted for about 48% of the total cost of hybrid rice seed (Janaiah and Hossain, 2000). Hybrid rice seed production requires about 340 man-days/ha, about 50 man-days/ha more than for ordinary rice cultivation. Hybrid seed production requires additional labor for extra farm operations such as thin and row planting, supplementary pollination, filling gaps, roguing, GA_3 application, and manual harvesting, threshing and cleaning.

Seed Production of Hybrids

Hybrid rice seed production technology is different from that for inbred rice seed production, and is more complex than the seed production for many other hybrid crops. Currently, the most popular male sterility system is the cytoplasmic genetic male sterility or CMS, popularly known as the three-line system. This utilizes three different lines, namely a cytoplasmic male sterile line (A line), a maintainer (B line), and a restorer (R line). Using this system for the commercial production of hybrid seeds involves two major steps. The multiplication of the A line (female parent) must be carried out before the production of hybrid (F_1) seed.

The other system, which is gaining popularity in China and Vietnam, and for which the efforts are on in India, involves a seed parent, which is an environmentally sensitive genetic male sterile (EGMS) or S line. This system requires neither a maintainer (B) nor a restorer (R) line. The seed parent can be multiplied just like any conventional inbred when planted in a fertility-inducing environment. When planted in a sterility-inducing environment, it becomes male sterile and may serve as a seed parent in producing F_1 seed of a hybrid. Any line found to form heterotic combinations with a particular S-line can be used as its pollen parent to produce a hybrid. Since this system involves two lines (TGMS or PGMS, and a pollen parent) to produce the hybrid seed, this system is also known as a two-line system.

Seed yield from a male sterile line used in hybrid rice seed production is a function of three factors: (1) yielding ability of the male sterile line, as determined by the yielding ability of its maintainer, the fertile counterpart, (II) the proportion of the male sterile line to the pollen parent, and (III) the out-crossing rate of the male sterile line. Improvement in any of these three factors can help to increase the seed yields of hybrid rice.

Extensive research has led to the identification of the following guidelines for successful hybrid rice seed production (Yuan, 1985; Mao, 1988; Virmani, 1994a):

- Selection of seed and pollen parents with synchronized time of anthesis
- Selection of seed parents with long, exserted stigma, longer duration, and wider angle of floret opening

- Selection of a pollen parent with a high percentage of residual pollen per anther after anther exsertion. High pollen shedding potential is attained by getting 2000–3000 spikelets/m^2 to bloom per hour during peak flowering period
- Synchronization of flowering time of the two parents by seeding them at different dates depending on their growth duration or estimated accumulated temperature requirements for initiation of flowering
- Use of optimum seed parent: pollen parent row ratio such that the ratio of spikelet number per unit area of seed parent and pollen parent is about 3.5:1
- Use of seed and pollen parents with small and horizontal flag leaves, or cutting of long and erect flag leaves
- Use of gibberellic acid (GA_3) to improve panicle exsertion and long duration of floret opening and stigma receptivity
- Planting of seed and pollen parent rows across the prevailing wind direction and use of supplementary pollination with a rope or stick when wind velocity is below 2.5 m/sec
- Selection of optimum time of flowering of parental lines in seed production plots

Large scale seed production is an important link between the researchers developing the hybrids and the rice farmers growing them. Success of hybrid rice technology primarily depends on purity, timely availability and affordability of hybrid seed to the farmers. Production of pure hybrid seed in a self-pollinated crop like rice, at affordable price is a highly skill oriented activity. Through extensive trials on different components like suitable locations, seasons, planting time, planting geometry, row ratios, GA_3 application and supplementary pollination etc. a package for hybrid seed production has been developed.

Success and sustenance of hybrid rice technology depend on efficient and economic seed production on large scale which is vital for popularization and large scale adoption of hybrids. However good a hybrid may be, unless it is feasible to produce the seed on large scale economically and efficiently, it will not reach large number of farmers. Cost effective seed production technology bridges the gap between technology generation and technology transfer.

Cultural Management

The hybrid rice seed production plot is given good cultural management. Weeds, diseases and insects are controlled, and the optimum fertilizer rate is applied. In case the panicle initiation stage of the two parental lines is found to occur at different times, they are given different application rates of nitrogen (urea), phosphate and potassium fertilizer, or draining/flooding the field with water, spray of chemicals to adjust the flowering date of the parental lines by 2–5 days.

If the parental lines have long, erect flag leaves which might obstruct the movement of pollen, the flag leaves are clipped by one-half or two-thirds of their length, one to two days before the initial heading. However, this practice has the negative effect that it may spread diseases such as bacterial blight and sheath blight through the wounds inflicted on the leaves. Besides, this practice also reduces photosynthetic activity. Parental lines possessing short narrow, flag leaves do not need to be clipped.

GA_3 is sprayed on parental lines which show poor panicle exsertion. Besides improving panicle exsertion, GA_3 also increases the stigma exsertion rate, as well as the time during which the spikelet is open and the stigma is receptive. Differential application of GA_3 also increases the relative height of the pollen parent over the seed parent. GA_3 is generally sprayed using a knapsack sprayer. Spraying begins when 5–10% of the rice plants are at the heading stage, and lasts for 2–4 consecutive days. Spraying is done on a calm sunny day, usually in the afternoon. It should not be carried out if rain is expected within 24 hours.

Hybrid rice seed producers in China use very high dosage of GA_3 (150–300 g/ha) to get high seed yields. However, outside China the high cost of GA_3 limits seed growers to using only 45–50 g/ha. The use of an ultra low volume (ULV) sprayer has been found to reduce GA_3 dosage to 15–20 g/ha, and give seed yields similar to those from spraying 45–50 g GA_3 from a knapsack sprayer.

The horizontal dispersal of pollen grains can be increased by shaking the panicles of the pollen parent by pulling a rope between two people, or stirring the canopy layer of the pollen parent with a bamboo stick (supplementary pollination). This should be done during the anthesis of the pollen parent. Either method may increase outcrossing, and hence seed yields. Care is taken that panicles are not broken at the neck. Roguing of off-types is carried out at maximum tillering, flowering and maturity stages, to prevent any cross-pollination between them and the true-to-type seed and pollen parent plants. Pollen and seed parents are harvested, threshed, dried and cleaned separately and carefully, to avoid any mechanical mixture of seed of parental lines.

Using the above practices, seed yields of 0.5 to 4 mt/ha (mean 1.3 mt/ha) have been obtained in the tropics. However, because of the vast experience of China's seed growers and specialized hybrid rice seed production, yields in China range from 1.5 mt/ha to 7.4 mt/ha (mean 2.5 mt/ha). In the late 1970's, average hybrid rice seed yield in Mainland China were 0.67 mt/ha. Currently, these have increased to 2.5 mt/ha. Based on the experimentation on seed technological aspects during the last decade in India, a package has been developed, a summarized version of which is given in Table 2.6.

Table 2.6. Optimum package for hybrid rice seed production and CMS multiplication

Activity	Particulars
Seed rate	"A" line or female parent : 15 kg/ha "B" or "R" line or male parent : 5 kg/ha
Nursery	Sparse seeding (15–20 g/m^2) to ensure multi-tillered (4–5) seedlings in 25 days
Row ratio	B:A ≡ 2:8 for CMS multiplication R:A ≡ 2:10 for hybrid seed production
Number of seedlings/hill	2 seedlings/hill for female parent 3 seedlings/hill for male parent
Spacing	Male : male = 30 cm; male: female = 20 cm Female : female = 15 cm; plant : plant = 15 cm or 10 cm
GA_3 application	60–90 g/ha/500 litres of water at 5–10% heading in two split doses on consecutive days
Supplementary pollination	Four to five times a day at peak anthesis within 30 minutes interval during flowering phase
Rouging	At vegetative stage–based on morphological characters of leaf and plant type At flowering–based on panicle characteristics At maturity–based on grain characteristics and percent seed set
Seed yield	1.5–2.5 t/ha

Initially the seed yields obtained were very low (0.3 to 0.5 t/ha), but with experience over the years, 1.5 to 2.5 t/ha average seed yields are being obtained now.

Quality Control for Hybrid Rice Seed

Quality control is vital to the widespread adoption of hybrid rice. Farmers who buy expensive hybrid seeds expect high quality. Quality control is carried out throughout the whole process of seed production. It includes the nucleus, breeder, foundation and certified seed production of both parental lines and F_1 hybrids.

Seed quality standards are met by intensive roguing of off-types (from seeding to harvesting) and careful handling during harvesting, threshing, drying, cleaning, processing, bagging and labeling. Grow-out tests, germination tests, seed vigor tests etc. are done before the seeds are sold commercially. There is a well-established national standard for hybrid rice.

Vegetative Propagation of Hybrid Rice

Vegetative propagation of rice is well known. Adult plants can be raised from seedlings, tillers, culm cuttings or by ratooning. Clonal propagation of rice was proposed in the early sixties to exploit hybrid vigor in rice

(Richharia, 1962). However, no serious attempt was made to develop this as a technology. Since the commercialization of hybrid rice technology in China in 1976, agronomists and farmers have tried using vegetative propagation in various ways to reduce the seed rate, and hence the seed cost, of commercial rice hybrids. These approaches include double transplanting and ratooning.

In double transplanting, hybrid seeds are sown densely in the seed bed. When the seedlings reach the 2–3 leaf stage, they are transplanted into a field nursery, using a wide spacing of 33 × 10 cm to induce more tillering. After 25–30 days, tillers are separated and transplanted into the main rice field. Although double transplanting can reduce the seed rate of hybrid rice significantly, it increases the labor cost. In subsistence farming, where family labor is readily available and there is no other opportunity cost, this method may be useful. However, in situations where labor is either not available or can earn more elsewhere, this approach may not be economical.

Ratooning of hybrid rice has been tried with varying degrees of success. Recently, successful vegetative propagation of rice has been achieved by splitting and transplanting the ratoon seedlings of an F_1 rice hybrid in Bangladesh. This approach would be practical with resource poor farmers who can use family labor but who cannot afford the full cost of hybrid seed.

Problems in Large Scale Hybrid Rice Seed Production

(a) **Maintenance of parental lines for three-line hybrids**: Parental lines are generally purified and maintained using conventional method of paired crossing. However, anther culture technique has also been used to purify the parents of few hybrid combinations in China (Zhang *et al.*, 1994). Purification approach essentially involved four steps:

(i) Growing the parental lines

(ii) Selection of typical plants and making crosses (back crossing and test crossing)

(iii) Evaluation of crosses in the identification nursery and multiplication of lines in the multiplication nursery separately planted at 30-day interval

(iv) Bulk multiplication of lines (breeder and foundation seed production).

(b) **Maintenance of parental lines for two line hybrids:** Virmani *et al.* (1997) have described a procedure based on field testing for maintaining the purity of and producing nucleus seed of TGMS lines in the tropics as:

Step I: Select about 100 typical plants from TGMS population grown under fertility inducting environment. Bag all the panicles. Select about 50 plants with the maximum fertility based on percent spikelet fertility.

Step II: Grow single plant progeny rows in sterility inducing environment and mark those that are completely sterile. Bulk the seed of these progenies kept in reserve from the nucleus seed in previous year.

Step III: Multiply the nucleus seed to produce breeder seed under strict isolation.

At Pantnagar, early wet season is ideal for complete male sterility (May- September) and utilized for identification of completely male sterile plants in segregating TGMS population or for hybrid seed production in isolation. Selected plants are allowed to continue for self seed reproduction during early winter (September- November) and the plants with maximum fertility are selected and bulked as nucleus seed.

(c) **Location for hybrid rice seed production:** At present large scale hybrid rice seed production is concentrated in only two districts of Andhra Pradesh, *viz.*, Karimanagar and Warangal. More than 90% of the hybrid seed is being produced in this region. With increase in demand for hybrid seed, additional areas are required for seed production. Area suitable for seed production in Karimnagar region has already reached saturation with almost 15,000 ha under seed production. Therefore, there is an urgent need to identify new areas suitable for large scale seed production in other states. A beginning has already been made by initiating large scale seed production in Nalgonda, Hanuman region of Andhra Pradesh; Raichur and Bellary districts of Karnataka; Salem and Coimbatore districts of Tamil Nadu; Kolhapur district of Maharashtra; Udham Singh Nagar and Nainital (Tarai) of Uttarakhand. Similarly, the possibility of undertaking seed production in late kharif season or in *boro*/summer season need to be explored in southern and eastern states.

Performance of public seed sector in hybrid rice seed production has not been encouraging so far, though many good public hybrids have been released. Seed production personnel in public sector have to be motivated to undertake large scale hybrid rice seed production. Hence the National Seeds Corporation, State Farms Corporation of India and the State Seeds Development Corporations have to be encouraged and provided all the needed facility and infrastructure to take up large scale hybrid seed production.

The time gap between harvest of seed at the end of *rabi* season and requirement of seed placement with the seed traders for *kharif* season cultivation, particularly for the northern India, is too short. To overcome this problem, the nurseries for seed production plots need to be seeded early during the second fortnight of October and the innovative molecular marker technology for detecting the purity of hybrid seed need to be adopted.

Strategies to Encourage On-farm Hybrid Seed Production

Progressive farmers for generations have been saving seed from healthy, better filled, disease free plants from parts of the field with a uniform stand after removal of visible off-types, followed by drying, cleaning and storing in reduced moisture conditions for next year planting or exchange with neighbour farmers. Such progressive farmers need to be trained with the hybrid rice seed production technology and be involved for seed production of public bred hybrids. Respective State governments should improve the rural based on-farm seed production system through:

- A clear policy statement on providing assistance for improving quality and yields of farm production of hybrid seed
- Identifying progressive farmers through local extension agents or other concerned government department officials or even NGOs, registering them as good seed multipliers, providing them basic training in hybrid seed production to increase unit yields and for quality control to maintain genetic purity
- Monitor their production and if found satisfactory declare them as good 'seed multipliers' and encourage neighbouring farmers to buy their seed requirements from them
- Ensuring that whenever a new adoptable hybrid variety is identified, a small quantity of its source seed (breeder seed/foundation seed) should be provided to the identified multipliers The distinguishing morphological characters of the variety and its production technology should be explained to them individually or in groups
- Giving the identified farmers instructions on quality monitoring through maintaining recommended isolation distances, timely rouging, harvesting at appropriate seed maturation phase, drying, seed treatment as a safeguard from storage grain pests, packing, if possible, in moisture proof containers, and storing under low humidity and temperature conditions
- Kisan melas or group meetings be organized at planted sites at seed maturation phase and farmers reactions should be ascertained. If found superior, farmers should be encouraged to undertake/increase production of the new variety until its diffusion through the system
- Regular technical support and frequent on-farm training of progressive farmers would maintain flow of quality hybrid seeds of new hybrids at reasonable cost and help increase national productivity, and
- Aiding farmers, who are sometimes in urgent need of cash, especially under natural calamity situations and are thus compelled to sell their seed. National seed policies should make provision for easy credit

availability at a reasonable interest rate to ensure that seed is not consumed. The other option is to develop a community fund at local level to purchase seed from such needy farmers with the option to return it or sell it within the community at a reasonable price

Popularization of Hybrid Varieties

Hybrid rice being a new technology for Indian agriculture, there is a need to create awareness about it among the Indian rice farmers for popularization of the hybrids. There is also a need to train farmers, farm women, seed growers, seed production personnel from public and private sector seed agencies and extension personnel on various aspects of hybrid rice technology, particularly on cultivation and seed production. Hence to create awareness about hybrid rice technology, Front Line Demonstrations (FLD) were conducted across the country in large number of compact blocks (Table 2.7). Yield advantage of 300 kg/ha to 2450 kg/ha was obtained during the period from 1999–2002.

Table 2.7. Compact block Front Line demonstrations conducted

State	Hybrids used	No. of demos conducted (1 ha each)	Yield advantage (kg/ha)
Maharashtra	Sahyadri, PHB 71, 6201	651	950–2450
Karnataka	KRH 2	550	700–1800
Andhra Pradesh	DRRH 1, APHR 2, PHB 71	475	300–1150
West Bengal	CNRH 3, KRH 2, PHB 71, 6201	410	500–1275
Tamil Nadu	CORH 2, ADTRH 1	330	400–0900
Goa	KRH 2, Sahyadri	300	900–2300
Uttaranchal	PSD 1, PHB 71	196	600–1350
Uttar Pradesh	NSD 2, PHB 71, 6201	186	900–2450
Orissa	DRRH 1, PHB 71, KRH 2	130	500–1300
Others (Haryana, Bihar, Gujarat, M.P., Tripura and Delhi)	PHB 71, DRRH 1, 6201	88	450–1475

At present, the popularization of hybrid rice is expected to get a boost through the implementation of the National Food Security Mission in different states. With the availability of seeds for new hybrids released for such NFSM areas, it is expected to enhance the popularization of hybrids through the minikit and FLD programs.

Experiencing success in imparting the knowledge and necessary skills for hybrid rice cultivation and hybrid seed production, appropriate trainings should be organized in a more intensive mode for farmers, farm women, seed growers, seed production personnel of public and private seed agencies, extension functionaries of State Departments of Agriculture, officials of

SAUs and NGOs etc. in the public institutions for popularization of the recently developed hybrids.

For providing exposure to the farmers and farm women on hybrid rice cultivation and hybrid seed production interstate farmers' visits were organized under the All India Coordinated Program. Such exposure visits were organized by different centres like Mandya, Maruteru, Coimbatore, Karjat, Karnal etc. The farmers from one state had an opportunity to visit another state under the program and see to learn for themselves the hybrid rice cultivation and hybrid seed production activities. This activity was very helpful not only to disseminate the technology but also strengthened the human resources in the farming community for popularization of the rice hybrids. Such activity is also expected to be intensified under the NFSM program of the country.

Marketing of Hybrid

In India and also the other Asian countries, seeds of improved varieties/ hybrids are produced and supplied through formal and informal seed systems. The formal seed system comprises of followings:

- The public sector in the form of National Seeds Corporation or the Provincial (State) Seed Corporations, *e.g.* in India, Pakistan, Bangladesh and Indonesia
- Co-operatives, mostly supported by government with active participation of progressive farmers in the region, *e.g.* Philippines and Japan
- Direct national government's involvement through departments of agriculture or extension, *e.g.* in Sri Lanka, Thailand
- Multinational or transnational companies
- The domestic private sector with or without its own R&D, and
- Multinational joint ventures, mostly with domestic private companies.

Seed supplied through the formal system is characterized by: planned production; some form of mechanized processing; named varieties (registered, released or notified); seed marketed in identified packages; a system of quality assurance to the buyers and crops in which it is inconvenient for the farmer to save their seeds. As a result there is a clear distinction between seed and grain.

The informal seed system provides for over 80 percent of seed needs in most developing countries. It comprises:

- NGO-supported seed multiplication and supply programmes;
- community groups

- farmers' cooperatives
- seed growers' associations, and
- farmer-to-farmer exchange.

The informal seed (supply) systems function in all or most developing countries of the region. It is characterized by lack of functional specialization, is heterogeneous in space and flexible in time. New varieties spread primarily through diffusion (farmer-to-farmer). The system develops naturally over time in response to farmers' demand for seed from sources other than their own. Hence, it has a low level of organization, lacks formal quality control as seed is multiplied without any generation control and is not subject to seed trade regulations.

Government policies in India encourage the public sector and fail to recognize the capabilities of the private sector. They offer hidden or direct subsidies usually to the public sector only. The private sector has to cover all costs in the sale price hence it cannot compete with the subsidized selling prices of the public sector. Recently, public-private sector partnership is being promoted by Govt. of India.

Regulatory Mechanism and Changing Scenario

Seed laws, Plant Variety Protection and Farmers' Rights implementation will not only accelerate national private sector growth, but will also encourage easy import and export of seeds. Government seed policies should be formulated to facilitate development of integrated national seed systems to meet farmer's demand for quality seeds, in needed quantities, at prices which cover all research, production and marketing costs, plus enough profit to lead the enterprise to self-reliance and steady growth.

Lack of private sector access to public germplasm and breeder seeds, including materials from International Agricultural Research Centers, is a regulatory barrier to the development of the private seed sector. National seed policy should not impede private sector participation in variety development by legally restricting the availability of germplasm, segregating lines, non-released varieties or breeding material. In fact, the policy should encourage exchange of elite genetic materials such as populations, germplasm and parent lines free, or against reasonable charges, from national and international research centers

Laws in many countries require compulsory variety registration and/or variety release and notification before seed production and sale. This deprives the farmers of that country superior genetic material which could increase their unit yields and national productivity. The preferred solution would be voluntary registration. If national governments insist on compulsory variety registration, it should be limited to a few major crops only. Automatic registration for varieties registered in a list of other countries, with similar agro-climatic conditions, within the region creating a multi-country variety list as in European Union countries is another good option.

Only certified seeds can be sold in certain countries. Proprietary hybrids, although may be superior, are not eligible for certification. This restricts their production and sale thus retarding private sector growth. A preferred solution is voluntary seed certification, leaving the choice of in-house quality control or quality control through a certification scheme to the producing company. The law should, however, ensure that the seed is sold in tamper-proof packages and is adequately labeled.

As regards the minimum seed and field standards for hybrid rice, it is desirable that realistic standards, based on research data from national seed technology research, are formulated. Mandatory requirement of distance isolation for seed production as of now should include as an alternative of using time or barrier isolation to encourage seed production by farmers at his field.

Subsidies on seed in India: Government purchases of hybrid rice seed from the public sector are often subsidized. Private sector sale of the same produce would fail to compete in the market and, therefore, the growth of hybrid rice extension and cultivation would be hindered. If subsidies, either direct or hidden, are made available to the public sector, they should also be made available to the private sector.

Seed promotion campaigns: Promotional campaigns by governments mainly focus on public bred material. They do not promote proprietary genetic material even if it is superior. In fact, government should promote rice hybrids that would be best suited to the farmers' needs. Prevailing grain prices in commercial markets at planting time often influence the user farmer's choice of a crop and variety. Use of improved seed coupled with associated scientific production technology results in increased productivity of crop varieties. Government has to fix proper support prices to purchase surplus grain, considering the cost of cultivation and reasonable profit, to sustain farmers' interest in planting seed of improved hybrid varieties.

Linkage with Public and Private Sector for Popularization

A good linkage of public and private sector is essential for the successful development and dissemination for popularization of hybrid rice technology in India. Public sector is strong in technology development and policy support whereas private and NGO sectors are strong in handling the seed production, marketing and technology transfer. The country needs to harness the strength of each sector to develop and popularize the technology. This would require a clear understanding among the various sectors on a mutually acceptable model for partnership. Virmani (2003) proposed the preferred scenario for partnership as follows:

Public Sector: Strategic and selected applied research (breeding, seed production, agronomic management), nucleus seed production and policy support.

Private Sector: Adaptive and selected adaptive research, commercial seed production (breeder, foundation and truthfully labeled and certified hybrid seed), processing and marketing.

NGO/Farmers' cooperative: On-farm demonstration and hybrid seed production technology transfer.

Since past few years, realization of such importance of public private partnership in large scale hybrid seed production has started in the form of individual MOU's signed between private seed companies and the public Institutions to go for non-exclusive contracts for the purpose of hybrid rice seed production. The seed thus produced is marketed through the private marketing system. Of late the importance and need to distribute the seeds of hybrids through the public network has been realized. However, the problems arising from the technical aspects of seed certification along with determination of seed cost for making them available to the farmers at a cheaper rate needs a feasible solution. Commercial production of the public bred hybrids would only then be successful.

Protection of Hybrids

Hybrid rice is only economically viable if planted for one cropping season. If hybrid rice is used as seed during the subsequent cropping season, the resulting plants will be non-uniform and yield will deteriorate. Hence, farmers are technologically restricted of any form of seed saving and would have to buy new seeds every cropping season. Farmers are further restricted to freely use hybrid rice in countries like India that have IPR legislations on seeds like PVPFR laws and/or PBR. Farmers are legally prohibited from creating new strains from hybrid rice varieties or from selling new strains due to provisions against use of essentially derived varieties.

This displaces the dynamic system that sustained rice farming in many communities for centuries. Farmers' traditional rights to save, re-use and exchange seeds, already recognized in the International Treaty on Plant Genetic Resources in Food and Agriculture (ITPGRFA) are not recognized in hybrid rice systems. In Vietnam, a study in 1999 warned that hybrid rice poses a great logistical challenge for the seed supply systems to annually renew all seed. A crop failure in the seed multiplication system can be disastrous for farmers as farmers cannot rely on harvested grains for seeds.

As it is, 35–45% of the total cultivated areas of Vietnam rely on farm-saved seeds and 8–18% relies on exchanged seeds.

Impact of Hybrid Rice Technology

An impact assessment study was made by two independent agencies, *viz.*, IIM, Ahmedabad and the Society for the Management of Agricultural Projects, Kanpur in the country for five states in the Northern India and three states of South India. The state of Maharashtra in Western region and West Bengal in Eastern region were also included in the study. The salient findings of the expert groups were as follows:

- Hybrids give an yield advantage of 1.0–1.5 t/ha in the northern states over the highest yield of inbred check.
- Additional net profit per hectare from hybrid rice cultivation ranged between Rs 2781/- and Rs 6291/-.
- The net profit by undertaking hybrid rice seed production was of about Rs 21,000/- per hectare as seed yield levels of 1.0 t/ha and the farmer gets seed price of Rs 50/- per kg
- Some millers are reluctant to offer the same price for hybrids as for inbreds in the States of UP and Bihar
- Prospects of large scale adoption of hybrid rice are bright if grain quality of hybrids is improved with proper policy and institutional interventions.

The impact assessment study in respect of MS, AP, TN, Karnataka and West Bengal States revealed the net profit by crop cultivation to be only Rs 1,250/- per ha in AP as compared to Rs 6,000/- in TN and Karnataka. Similarly, the net profit in hybrid seed production was much lower in AP which was only Rs 7,500/- as compared to Rs 30,000/- per hectare in Karnataka. Hybrid seed production generates additional employment of 65 man days/ha.

Benefit-Cost Ratio

Adoption of the package of hybrid seed production technology could yield on an average about 2.0 t/ha hybrid seed yield on a large scale. At the average procurement price of Rs 30–40 per kg, the gross returns are in the range of Rs 60,000/- to Rs 80,000/- per hectare with the cost of production being Rs 25,000 to Rs 30,000/-. Thus the net profit from hybrid seed production would be Rs 35,000 to Rs 50,000/- per hectare. Honnaiah (2003) reported cost of hybrid seed production to be Rs 24,188/ha as against Rs 20,815/- for an inbred. In terms of returns ha^{-1}, hybrid seed production gave return of Rs 44,012/- as against Rs 25,125/- giving an additional net profit of about Rs 20,000/- per ha from hybrid seed production (Tables 2.8 and 2.9).

Table 2.8. Cost of cultivation for the hybrid rice and improved rice variety seed production (Honnaiah, 2003)

Input	Hybrid rice		Improved varieties (Jaya/IR 64)		% increase over check
	Rs	%	Rs	%	
Seeds : A line	1,875	6.90	7,125	5.40	106.67
B line	450	1.66	–	–	–
FYM	2,500	9.20	2,500	12.01	0.00
Fertilizer and PP chemicals	4,108	15.11	4,300	20.66	- 4.47
GA_3	2,400	8.80	–	–	0.00
Human and bullock labour	14,605	53.72	10,870	52.32	34.36
Miscellaneous	1,250	4.59	2,020	9.70	- 38.12
Total cost	24,188	100.00	20,815	100.00	

Table 2.9. Returns in hybrid rice seed production and seed production of improved varieties per ha

Particulars	Hybrid rice seed production			Improved varieties (Jaya/IR 64)		
	Quantity (kg)	Rate (Rs)	Value (Rs)	Quantity (kg)	Rate (Rs)	Value (Rs)
Seed yield : A line	1570	40	62800	5000	8	40000
R line	1180	5	5900	–	–	–
Straw	5T	500	2500	7.50T	400	3000
Gross income (Rs)		71200			43000	
Cost of cultivation (Rs)		27188			20815	
Net return (Rs)		44012			22185	
Cost: benefit ratio		1:1.62			1:1.1	
Cost of seed production (Rs/kg)		17.32			4.14	

It is thus a profitable venture for the seed growers in the states of Uttaranchal, A.P., Karnataka and Maharashtra who have benefited by undertaking the hybrid rice seed production. Also, the hybrid rice seed production generated additional employment for 60–80 man-days per hectare in rural India.

Major Constraints of Hybrid Rice Technology

Over the past decade, hybrid rice promotion has been beset with problems that hamper continuous adoption by rice farmers:

Inconsistent yield advantage over inbreds: Contrary to claims that it poses 20–30% yield advantage over conventional inbred rice varieties, hybrid rice performance is location-specific and is ideal in sub-tropical and sub-temperate zones. Experiences in some Asian countries attest the fact:

- China–only 15–16% yield advantage over inbred counterparts in Jiangsu Province in 1984. Yield of hybrid rice varieties developed from CMS systems are reportedly declining.
- Vietnam–hybrid rice cultivation is concentrated in the Northern Vietnam during the winter months.

Hybrid seeds are vulnerable to seed-borne diseases: Hybrid rice seeds, as compared to inbred rice seeds, are partially opened, discolored, half-filled and lightweight. Being partially opened, it is more susceptible to seed borne insect pests and diseases, especially under humid tropical conditions. In the Philippines, where such conditions prevail, hybrid rice seeds need cold dry storage facilities, which proved to be more costly for seed growers to maintain. In Northwestern India and China, CMS lines in hybrid rice seed production plots were reportedly found to have a higher incidence of seed-borne diseases, *i.e. paddy bunt* and *false smut*.

Inferior grain quality: Despite long slender grains, there is a lower price for hybrid rice grains than premium inbred varieties in the market. Lower market price for hybrid rice grains in China, Vietnam, India and Bangladesh were reportedly due to its inferior quality. In Thailand, the lack of demand for hybrid rice is mainly due to market preference for Jasmine Rice. In the Philippines, despite higher premium of hybrid rice compared with ordinary rice, milled hybrid rice is sold as ordinary rice because of its low head rice recovery.

Costly seeds and technology: Hybrid rice seed production is labour, input and knowledge intensive. It requires additional 50 man-days/ha than normal mechanized rice cultivation. It is highly dependent on the use of gibberellic acid, a growth regulator needed to synchronize the flowering of the hybrid seed parents. Outside of China, gibberellic acid is expensive (more than $1.00 per gram). In India, the average hybrid seed cost is Rs 1,500 per hectare while inbred varieties cost Rs 300–400/- per ha only.

Irrigation and transplanting of rice seedlings warrant successful hybrid rice cultivation. In China, almost 100% of the rice area is irrigated and transplanting is practiced in 97% of the rice area. Dramatic change of cultivation and farming systems is required if hybrid rice is to planted in rainfed and upland conditions.

Threat of genetic erosion: Loss of genetic diversity has become a major concern in India and other Asian countries where the modern varieties have become more vulnerable to pests and diseases. Around 91% of the post IR-8 varieties are genetically related and have a single parental source which can be traced to a common cytoplasm called *Cina*. With hybrid rice, problem of cytoplasmic uniformity is graver where almost 100% of the total area planted with hybrid rice in India came from a single cytoplasm source–the WA cytoplasm utilized in the CMS system. Even in China, despite efforts to diversify the CMS lines in rice, about 88% of the commercial hybrids carry WA cytoplasm. This cytoplasmic uniformity could result in disease epidemics similar to the Southern corn blight outbreak that wiped out US hybrid maize due to the use of a single cytoplasm.

Possible displacement of the role of women-farmers: In India, the women play a key role in all activities involving farm-seed saving, *i.e.* handling, harvesting to raising seedlings for the succeeding cropping season. However, their role will be displaced if India continues to promote hybrid rice on a massive scale as in Vietnam.

Opening the flood gate to corporate control of the rice seed sector: Hybrid rice has already set the stage for 'privatization' of the hybrid rice seed sector despite the key role played by public research institutions in the initial stages of hybrid rice programs. Already, there is an increasing trend of private seed companies involved in hybrid rice promotions across Asia. More than 95% of the hybrid seed production in the country during 2006 was in the private sector.

CONCLUSION

Significant cross pollination in the cytoplasmic male sterile lines of otherwise naturally self-pollinated crop of rice has made it possible to exploit this character for the purpose of realization of heterosis and thereby boost the national productivity level. Prospects of exploitation of inter sub-specific heterosis (*indica / tropical japonicas*) using thermo-sensitive genic male sterility (two-line breeding) with genes for wider compatibility has been explored with most encouraging results from two-line hybrids developed and tested in multilocation trials under All India Coordinated Rice Improvement Project. However, apomixis is the ultimate genetic tool to develop true breeding hybrids and facilitate commercial exploitation of heterosis by even resource poor farmers of the developing world. So far, there is no confirmed report of apomixis in rice, however, research is in progress in different institutes to discover, induce and/or genetically engineer such character in rice. Also, the prospect of vegetative propagation of rice hybrids is being explored and it is expected that the impact of such propagation of the planting materials would be felt during the next few years. Public–private partnership in large scale hybrid seed production has to be given a boost for universal adoption of the technology.

REFERENCES

Dwivedi, D.K., Pandey, M.P., Pandey, S.K. and Rongbai, L. 1998. Heterosis in inter- and intra-subspecific crosses over three environments in rice. *Euphytica* **99**: 155–65.

Dwivedi, D.K., Pandey, M.P., Pandey, S.K. and Rongbai, L. 1999. Studies on screening and genetics of wide compatibility in rice (*Oryza sativa* L.). *Ind. J. Genet.*, **59(3)**: 1–14.

Janaiah, A. and M. Hussain. 2000. Hybrid rice for food security in the tropics: An evaluation of farm-level experiences in India. Paper presented at the 3rd International Crop Science Congress held in Hamburg, Germany, 17–22 August 2000.

Mao, C.S. 1988. Hybrid rice seed production in China. *In*: Seed Health. International Rice Research Institute, Los Banos, Laguna, Philippines. pp. 277–82.

Pandey, M.P. 2004 Seed production in hybrid rice. *In*: Proc. National Symposium on "Harnessing Heterosis in Crop Plants" held at IIVR, Varanasi, March 13–15, 2005. 57–64.

Pandey, M.P. 2004. Technology of hybrid rice seed production and public-private partnership for large scale adoption. *In*: Souvenir "SAARC Rice Expo 2004" organized by Maharashtra Chamber of Commerce and Industry and Rice Exporters Association of Pakistan, held at World Trade Centre, Mumbai, Dec, 2004. 68–72.

Pandey, M.P. 2006. Researchable issues for the development of hybrid rice. *In*: Group Meeting on "Development of Hybrids for Rainfed Shallow Regions of Orissa, Bihar & West Bengal held at CRRI, Cuttack, August 5, 2006. 6–10.

Pandey, M.P. 2006. Hybrid rice seed production-Public sector experience and constraints. *In*: Group meeting on "Development of hybrids for rainfed shallow regions of Orissa, Bihar & West Bengal held at CRRI, Cuttack, August 5, 2006, 33–39.

Pandey, M.P., Ahmed Ilyas and Ramesha, M.S. 2001. Strategies for large scale seed production of hybrid rice in India. *In*: the National Seminar on "Rice Production Management" organized by the Ministry of Agriculture, Hyderabad, Sept. 12–13, 2001. 9.

Pandey, M.P., Singh, J.P., Shukla, S.K., Singh, P.N. and Dwivedi, D.K., 2002. Study on technique for economization of seed production cost in hybrid rice. Abstract in the 4th International Symposium on Hybrid Rice held at Hanoi, Vietnam, May 14–17, 2002.

Rao, R.N. and Patnaik, S.S.C. 2006. Current status of hybrid rice research and development for shallow low lands. *In*: Development of Hybrids for Rainfed Shallow Regions of Orissa, Bihar & West Bengal held at CRRI, Cuttack, August 5, 2006. 20–24.

Richharia, R.H. 1962. Clonal propagation as a practical means of exploiting hybrid vigor in rice. *Nature* (London) **194**: 598.

Viraktamath, B.C. 2006. Current status of hybrid rice in India. *In*: Group meeting on "Development of Hybrids for Rainfed Shallow Regions of Orissa, Bihar & West Bengal held at CRRI, Cuttack, August 5, 2006. 1–5.

Virmani, S.S. 1994. Heterosis and hybrid rice breeding. *In*: Monographs on Theoretical and Applied Genetics 22, S.S. Virmani (*Ed.*). Springer-Verlag.

Virmani, S.S., Viraktamath, B.C., Casal, C.L., Toledo, Lopez, M.T. and Monalo, J.O. 1997 Hybrid Rice breeding manual, Manila (Philippines): International Rice Research Institute, 151P

Yuan, L.P. 1985. A Concise Course in Hybrid Rice. Hunan (China): Hunan Science and Technology Press. 168 pp. Strategies for commercialization of public bred hybrids in India.

3

Single Cross Hybrid Seed Production in Maize

Sain Dass[*]

ABSTRACT

Maize is one of the major cereal crops with wide adaptability to diverse agro-climatic conditions around the world. Inadequate supply of hybrid seeds during growing season forces farmers to use poor quality seed causing 30–40% reduction in yield. Availability of quality seed is the key for growth of production and productivity. It can be achieved by region specific seed production programme of single cross hybrid with the concept of "Seed Village". Seed production technology, advantages and economics of single cross hybrids are discussed in detail.

Key words: Economics, maize hybrid, seed production, seed village, single cross hybrid

Maize (*Zea mays* L.) is one of the major cereal crops with wide adaptability to diverse agro-climatic conditions around the world. It is the third most important crop in India after rice and wheat that occupied about 8.0 million hectares having production of 16.78 million tones with an average productivity of more than 2.0 tonnes ha^{-1} (Fig 3.1). The growth rate of area, production and productivity of maize during last decade has been quite impressive with corresponding values of 2.2, 4.2 and 1.96% that shows a commendable response of the producers and the industry. The satisfactory growth in maize production is attributed to development of hybrids with high yield potential.

Being a highly cross-pollinated crop, maize requires a high percentage of seed replacement every season as compared to self-pollinated crops. Inadequate supply of hybrid seeds during growing season forces farmers to use poor quality seed causing 30–40% reduction in yield as the hybrid seed production is mainly confined to small pocket of the country, which is not available to farmers of other regions. Availability of quality seed is the key for growth of production and productivity and to cover 8 million hectares,

* Director, Directorate of Maize Research, ICAR, New Delhi - 110 012.

about 2.0 lakh tones of quality seed is required. Presently nearly 40% maize area is covered under hybrids of which the share of single cross hybrids (SCH) is 5–7% and the remaining 33–35% is under double cross or three-way cross hybrids which are less productive than the SCH.

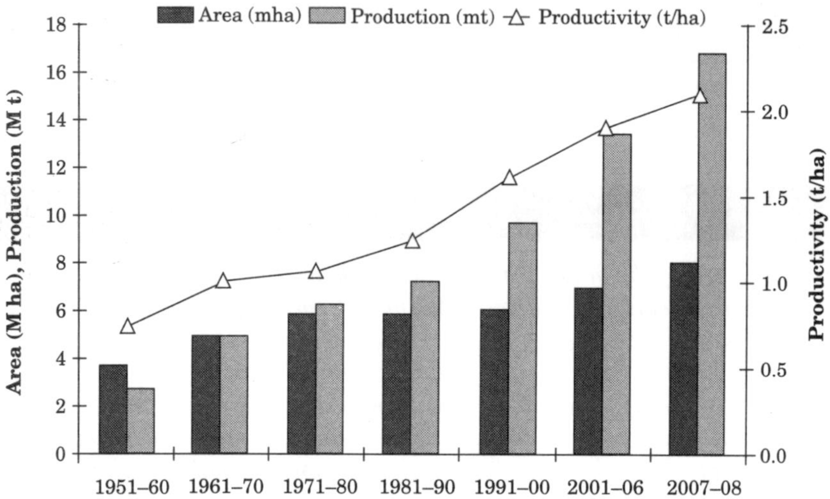

Figure 3.1. Trends of area, production and productivity of maize in India

Merits of Region Specific Seed Production/Seed Village

To meet the increasing demand of quality seed across the country, the seed production program should be region specific with a "Seed Village" concept that facilitates in:

- Overcoming isolation problem
- Ease in monitoring
- Maintaining genetic purity of the seed
- Reduction in cost of production
- Timely availability of seed
- Reduction in cost of transportation
- Improve farm profitability
- Employment generation to rural masses
- Earning foreign exchange as the cost of single cross hybrid (SCH) maize seed production is lowest in the developing and under developed world due to low cost of cultivation because of cheaper labour.

Advantages of Single Cross Hybrids

- High acceptability among the farmers
- Require only two parents

- Require less number of isolations, *i.e.* only three isolations
- Easy to multiply and involve less labour
- Low cost of production
- Tolerant to biotic and abiotic stresses
- Easy to market because of uniformity & high productivity
- Provide employment opportunity to their agriculture graduates.

Prerequisites of Single Cross Hybrid Seed Production

- Proper site selection-avoid the sites where the preceding crop was maize
- Fertile and quality land with good quality and assured irrigation
- Productive, uniform and genetically diverse inbred lines
- Proper isolation distance
- Technology (package of practices) for hybrid seed production
- Technically experienced manpower
- Stress-free season.

Characteristics of Good Seed Parent (Female)

- Productive
- Strong
- Long cobs with complete exertion
- Low cob placement
- Lax tassel
- Shorter anthesis silking interval (ASI)
- Nutrient responsive
- Stay green trait
- Erect leaves
- Resistant/tolerant to biotic and abiotic stresses
- Strong root system.

Characteristics of Good Pollen Parent (Male)

- Lax tassel, long main branch with few secondary branches

- Long duration of pollen shedding
- Taller than female parent
- Attractive grain colour
- Strong plant resistant to lodging with better root system
- High yield potential
- Resistant/tolerant to biotic and abiotic stresses

Important Considerations for Hybrid Seed Production

Isolation distance

If possible, the hybrid seed production should be taken either in the area where no other maize variety is planted nearby the seed production plot or at least 400–500 metre distance is required between two maize genotypes to maintain genetic purity. This is mainly possible under "Seed Village Concept" by taking seed production of one hybrid in one village which is most convenient for producing quality seed.

Male : female ratio

The male: female ratio depends on (a) the pollen shedding potential of male parent, and (b) male : female synchrony for better seed setting the flowering of the female should be earlier than the male or male pollen dehiscence should coincide with female silking. In general, the male : female ratio should be 1:2:1 :2 :1 or 1:3:1:3:1, or 1:4:1:4:1.

How to Ensure Male: Female Synchrony?

Male : female synchrony can be brought about by:

- Staggered planting of male and female lines
- Manipulation of plant distance by spaced and narrow planting
- Irrigation along with fertilizer appli-cation
- Application of FYM in either male or female to induce earliness and vigor.

Technology for Hybrid Seed Production

Seed production site: Seed production should be taken in well drained, weed and disease free soil and preferably in the fields where the preceding crop was not maize to minimize rouging and maintain the genetic purity.

Time of sowing: Appropriate time of sowing is very important for better crop establishment. For most part of India, the first week of July during *kharif* and first week of November during *rabi* are the optimum time of sowing to avoid flowering under heavy rains during *kharif* and low temperature during winter. For better pollination the rain and low temperature should not coincide with flowering. Rains during flowering wash the pollen in *kharif* and low temperature during winter causes plant mortality and kills anthers.

Method of sowing and layout: It is desirable to plant the crop on raised beds (ridges). Sowing should be done on the southern side of the east-west ridges, which helps in good germination. Planting should be done at proper spacing. Optimum row and plant spacing should be kept at 60 and 20 cm, respectively. This spacing will ease the movement in the field for rouging and removal of tassels. Proper spacing also helps in improving the test weight. Identification labels/tags should be put on the male and female lines to distinguish between them.

Seed rate: The seed rate depends on (a) seed size/test weight of male and female seed, (b) plant type as in case of erect plant higher seed rate is required to accommodate large number of plants per unit area than in spreading plant types, and (c) male : female ratio. In general the optimum seed rate is 15 kg/ha for female and 10 kg/ha for male with at least 80% germination.

Seed treatment: Seeds should be treated with fungicides and insecticides before sowing to protect it from seed and soil borne diseases and some insect-pests.

- Bavistin + Captan in 1: 1 ratio @ 2 g/kg seed for TLB, BLSB, MLB etc.
- Apran 35 SD @ 4 g/kg seed for BSMD
- Captan 2.5 g/kg for Pythium Stalk Rot
- Imidachlorpit @ 4 g/kg for termite and shoot fly.

Nutrient management: In general, inbreds are slow growing and weak in nature and have poor nutrient uptake efficiency. Therefore, inbreds require high fertility as compare to hybrids. It is desirable to apply FYM @ 15 t ha^{-1} 15 days prior to seeding. The fertilizer nutrients required for inbreds are N 180–200 kg, P_2O_5 80 kg, K_2O 80 kg and $ZnSO_4$ 25 kg ha^{-1}.

Full dose of phosphorus, potash and zinc and 10% N should be applied as basal. The remaining dose of nitrogen should be applied in four splits as per details given below to avoid losses and meet the requirement throughout the crop cycle:

1. 20% N at 4 leaf stage
2. 30% N at 8 leaf stage
3. 30% N at flowering stage
4. 10% N at grain filling.

Water management: The first irrigation should be applied very carefully. Water should not overflow on the ridges. The irrigation should be applied in furrows up to 2/3 height of the ridges. Irrigation should be given as and when required by the crop depending on the rains and moisture holding capacity of the soil.

Light and frequent irrigations are desirable for inbred parents. Young seedlings, at knee height stage, flowering and grain filling and 10 days after grain filling are the most sensitive stages for water stress for inbreds and irrigation should be ensured at these stages.

Weed management: Weeds are serious problem in maize, particularly in wet (monsoon) season that competes with maize for nutrient and causes yield loss up to 35%. Therefore, timely weed management is needed for achieving higher yield. Atrazine being a selective and broad-spectrum herbicide in maize checks the emergence of both broad leaves and most of the grasses. Pre-emergence application of atrazine @ of 1.0–1.5 kg a.i ha^{-1} in 600 litre water is effective way for control of weeds. While spraying, the person who is doing the spray should move backward so that the atrazine film on the soil surface is not disturbed. Preferably, three nozzle booms may be used for proper ground coverage and saving time. One to two hoeings are recommended for aeration and uprooting of the remaining weeds. While hoeing, the person should move backward to avoid compaction and ensure better aeration.

Protection from serious insect pests: Stem borer is a serious problem in maize. It can be controlled by 1–2 sprays of Carboryl or Endosulfan after 10 days and 20 days of germination. Spraying should be done in the central whorl of plants.

Ist spray: 500 g Carboryl or 625 ml Endosulfan (35EC) in 500 litre water.

IInd spray: 750 g Carboryl or 900 ml Endosulfan (35EC) in 700–800 litre water, if required.

Removal of Off-type Plants and Thinning

(i) At early stage, *i.e.* 12–15 days after sowing, off-type plants and excess plants should be removed and proper plant to plant distance of 20–25 cm should be maintained to provide an equal opportunity for each plant to grow,

(ii) At knee high stage, and

(iii) At flowering, *i.e.* before anthesis.

Dissimilar plants should be removed from the male and female lines to maintain the genetic purity of seed. Dissimilar tassel bearing male plants should also be removed.

Earthing up: One day prior to earthing up, the third split of nitrogen should be applied followed by hoeing. On the next day earthing up operation should be completed with movement of the person in the backward direction. This operation should be done before tasseling stage to save the crop from lodging.

Detasseling: Detasseling in female plants should be done before anthesis. It should be practiced row-wise. One person should follow to monitor each row to check that no part of the tassel is left inside. The process of detasseling should continue for 8–10 days. While detasseling, leaf should not be removed which will otherwise reduce the photosynthesis. It has been observed that the removal of 1 to 3 leaves along with tassel reduces 5–15% yield. The removed tassel should not be thrown in the field but fed to the cattle as it is a nutritive fodder.

Harvesting: The male parent should be harvested before the female and should be kept separately. Optimum moisture content in grain at harvesting should be around 20%. The harvested cobs should spread evenly instead of making heap.

Stages of Crop Inspection

- At the time of sowing: to monitor the land, isolation distance, planting ratio of male : female, proper sowing time and seed treatment
- During pre-flowering/vegetative stage: to verify the rouging and removal of off type plants
- During flowering stage: to check disease and pest infestation
- During post-flowering and pre-harvest stage: to remove the late and diseased plants
- Harvesting time: to decide the proper time of harvesting.

Post-Harvest Management

Drying and sorting of seed parent cobs: The drying of the cobs should not be done either on the kuccha or pucca flour, rather it should be dried on tarpoline sheets to avoid seed injury and during night the cobs should be kept covered. To maintain purity, dissimilar, diseased and pest infested cobs should be removed before shelling. The female cobs should be dried upto 13–14% moisture content before shelling.

Shelling: Female parent should be shelled earlier than male to avoid mechanical mixture. Shelling can be done manually or by power operated maize sheller.

Seed processing: All undersize, broken, damaged etc. seeds should be removed for maintaining the quality of seed.

Storage and marketing: Seed drying should be done till the moisture content of the seed is reduced to 8% and it should be kept in aerated jute bags. Seed should be stored at cool and dry place preferably in cold storage. Poor storage conditions will lead to loss of vigour and poor germination. Marketing should be done with specifications and standards.

Economics of Seed Production of Single Cross Hybrids

The economics of seed production depends on (a) productivity, (b) genetic purity, (c) effective seed production technology, and (d) management practices. Under Indian conditions, the general economics of seed production of single cross hybrids is as under:

Operations	Amount	
	Rs/ha	**US $/ha**
1. **Field preparation and sowing:** (Harrowing, tiller planking, and ridges making)	3,500	83
2. **Input cost**: (Seeds, fertilizers, pesticides, herbicides, fungicide)	8,500	202
3. **Crop and field management:** (Hoeing, weeding, earthing, irrigation, fertilizer, herbicide, pesticide application, watch & ward etc.)	10,000	238
4. **Detasseling and rouging**	1,060	25
5. **Harvesting**	1,650	39
6. **Drying and sorting of cobs**	450	11
7. **Shelling and seed drying**	1,350	32
8. **Land rent for six months**	18,000	428
TOTAL	**44,510**	**1,060**
Seed yield @ 1.5 to 2 tons/ha		
Cost of F_1 seed @ Rs 45/kg		
Gross income Rs 67,500 to 90,000/-		
Cost of production per kg of seed*	**Rs 22.25/kg**	**$ 0.53/kg**

* *The cost does not include processing, packaging, transportation etc.*

The major share of cost involved in seed production is land rent (Fig 3.2) followed by input costs and crop management practices.

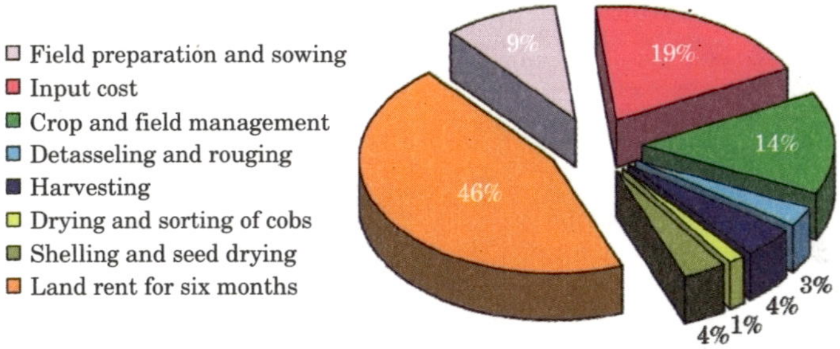

Figure 3.2. Share of different inputs in cost of seed production

4 Strategies for Commercialization of Public - Bred Hybrids of Maize in India

H.S. Gupta and V. Mahajan*

ABSTRACT

VPKAS, Almora, released many hybrids, composites and varieties of speciality maize with the distinction of developing first double cross hybrid in the country. The centre converted normal maize hybrid to quality protein maize hybrid by deployment of marker assisted selection that shorten the breeding cycle. Exhaustive and diverse germlplasm is maintained and documented with free exchange alongwith development of a large number of inbred lines and heterotic pools. Economization of seed production of single cross hybrids by narrow-based seed parents, marker aided selection for quality protein maize, molecular breeding scheme and other related aspects are discussed in detail.

Key words: Commercialization, germplasm, hybrid maize, marker aided selection, quality protein maize, specialty corn

The Vivekananda Parvatiya Krishi Anusandhan Sansthan, Almora, has a distinction of having developed first double cross hybrid (VL *Makka* 54) in the country even before the inception of AICRP on Maize. Thereafter, sustained research helped in development and release of **thirteen more hybrids** and **eight composites,** which are recommended for the northwestern hills as well as other parts of the country. Recently, we developed a plan for increasing hybrid seed production of single cross hybrids using narrow-based seed parent. This approach led to the development and release of 'Vivek Maize Hybrid 15' and 'Vivek Maize Hybrid 17' by CSCSNRV in 2004. The seed yield of this hybrid increased by 60–70% over the best hybrid. In addition, we have developed a number of populations and inbred lines that were shared with other maize researchers in the country. The institute is regularly producing and

* Vivekananda Parvatiya Krishi Anusandhan Sansthan, (Indian Council of Agricultural Research), Almora 263601.
Corresponding author: Email: vpkas@nic.in

supplying composites and hybrids to different organizations. Deployment of marker assisted selection has helped in shortening the breeding cycle and led to conversion of normal maize hybrid to quality protein maize hybrid for the first time. This novel technology has aided in improving protein quality of a popular hybrid, which will, in turn, help in alleviating malnutrition in monogastric animals, including human being.

INTRODUCTION

Maize is an important cereal crop of north-western Himalayan states and occupies an area of 6,48,400 ha area (90% area is rainfed) with an average productivity of 1604 kg/ha which is *at par* with the national productivity of 1938 kg/ha in 2005–06. Development of superior maize hybrids is a pre-requisite for increasing its production and productivity in the country. Vivekananda Parvatiya Krishi Anusandhan Sansthan (VPKAS) has released 14 hybrids including a QPM hybrid (Vivek QPM 9) and 8 composites that are recommended for commercial cultivation in the different parts of the country.

Historical Background

Realizing the importance of maize in hill agriculture, research work was initiated at this institute as early as 1948 by late Professor Boshi Sen, the Founder Director, by introducing various dent double cross hybrids from USA. Of these, US-13 and NC-27 yielded 95% more than the best check of that time, KT-41. However due to dent nature, low seed setting and susceptibility of their parental lines to leaf blight (*Helminthosporium turcicum*) these hybrids did not pick up well. Efforts to overcome these drawbacks, under ICAR funded scheme in 1952, were intensified with the inception of All India Coordinated Maize Improvement Project in 1957. These efforts led to the release of a late maturing (115–120 days) orange flint double cross hybrid VL 54 having wider adaptability and resistance against leaf blight (*H. turcicum*) by the Central Varietal Release Committee [now renamed as Central sub-committee on Crop Standards, Notification and Release of Varieties (CSCSNRV)] in 1962 for the Himalayan hills.

Thereafter, a number of hybrids were developed and tested over a period of sixty years at different locations. Based on agro-climatic conditions and altitude, hills can broadly be grouped into lower (< 1,000 m), mid (1,000–1,500 m) and higher hills (> 1,500 m). With the increase in altitude, maturity period of a variety prolongs which in turn may make a variety unsuitable for taking two crops in a year, and requires careful selection of varieties for different altitudes. Due to short growing period rain is available to the *kharif* crops in hills where generally the varieties, which mature within

90–95 days, are suited to fit in sequential cropping. Hence, early to early-medium duration maize cultivars get top preference in these areas. Considering the short growing period and high cropping intensity in hills, the emphasis has been on the development of early duration genotypes with high yield potential and resistance to prevailing diseases in general and *turcicum* leaf blight in particular. The accomplishments in maize research are presented hereunder.

Maize Research at VPKAS, Almora

Research work to increase yield potential of the one of the most important cereal crop of the country through exploitation of hybrid vigour was initiated at VPKAS as early as 1948 by introducing seed of three double cross dent hybrids and one improved sweet corn variety Golden cross Bantam from USA. The results obtained from these hybrids were quite encouraging and in 1949 seeds of sixteen new hybrids strains were introduced. They were tested during 1949 along with four top crosses made at Almora during 1948. During 1949 work on hybrid maize was also started at IARI, New Delhi. Multilocational trials were conducted at 17 locations throughout the country. Exceptionally high yields were recorded from the hybrids and the most promising cultures identified were US 13 and NC 27 among the VPKAS introductions and Dixie 11 and Dixie 22 among the IARI introductions. These hybrids yielded almost double than KT 41 the best check for the state of U.P. at that time. Seeds of the parental lines of the most promising hybrids were introduced by VPKAS (Sen, 1953) and seeds of hybrid US 13 were multiplied. However, despite high yields, owing to their highly dent grains, which were not relished by the consumers, susceptibility of their parental lines to turcicum leaf blight and poor seed yields these hybrids did not become popular with the farmers. Research work on maize improvement continued at VPKAS (Vivekananda Laboratory) under an ICAR scheme from 1952 onwards and the drawbacks of these US hybrids were removed to a great extent by utilizing the established inbred lines from USA, South Africa and other regions of the world. Several hybrids with round flint grains were developed till 1955, before inception of All India Coordinated Research Project in Maize in 1957. Maize research was further intensified as a part of AICRP.

Medium maturity strains: Several hybrids with round grains were developed till 1955. After three years of evaluation at Almora, three best among these, *viz.,* VL 23, VL 45 and VL 54 were evaluated at different centres of AICMIP. Of these, VL 54 was released by the CVRC in 1962 for general cultivation in the entire Himalayan region, including the foot hills. VL Makka 54 was a medium duration, yellow flint double cross hybrid combining higher yield potential and high degree of resistance to *turcicum* and *maydis* leaf blights. Another significant achievement in the development of medium duration varieties is the release of white flint grain double cross hybrid HIM 128 (VL 43) by CVRC in 1979 for cultivation in the hills. HIM 128 combined higher yield potential with a very high degree of resistance to the two most important diseases of maize in the hills, *viz., turcicum* and *maydis* leaf blight.

Early maturity strains: Medium duration hybrids offered high stability in production at lower elevation, however, they were found unsuitable for cultivation in the higher hills owing to their maturity duration of 120 days or more. Moreover, even in low altitudes, medium duration hybrids were found suitable only in the irrigated valleys and did not fit well to the cropping system under rainfed conditions, especially the areas growing two crops every year. Taking all these points into consideration a programme for the development of short duration hybrids was started in the early sixties. Attempts were made to introduce earliness in the high yielding background from the short duration local varieties and exotic germplasm. A number of hybrids with 95–100 days maturity period were developed and ultimately VL Makka 42 with orange flint grain was released for cultivation in U.P. hills by U.P. SVRC in 1972. It is highly adapted for cultivation above 1700 m amsl. During 1973, an early maturing composite VL Makka 1 was released for cultivation in U.P. hills by the SVRC.

Although a programme for the development of composites was initiated in 1966, it was in 1973 when the work on the synthesis of short duration composite was made the main objective of maize improvement programme at VPKAS. Work on the development of composites was intensified, with a virtual stoppage of research on the development of hybrids. This resulted in the development and release of several promising strains like VL Makka 16 (SVRC, 1980), VL Makka 41 (SVRC, 1986), VL Makka 88 (CVRC, 1988), Vivek Sankul 11 (SVRC, 2003) and Vivek Sankul 31 (SVRC, 2007). All these yellow composites combine earliness with superior yield and high degree of tolerance to turcicum and maydis leaf blight. The project on Promotion of Research and Development Efforts on Hybrids in Selected Crops was started in 1989. Since then, the institute has developed and released one double top cross hybrid, HIM 129 (CVRC, 1997), seven single cross hybrids, *viz.*, Vivek Hybrid 4 (CVRC, 1999), Vivek Hybrid 5 (SVRC, 1999), Vivek Hybrid 9 (CVRC, 2000), Vivek Hybrid 21 (CVRC, 2006), Vivek Hybrid 23 (CVRC, 2006), Vivek Hybrid 25 (CVRC, 2007), and Vivek Hybrid 27 (CVRC, 2007) and two top cross hybrids, namely, Vivek Hybrid 15 (CVRC, 2004) and Vivek Hybrid 17 (CVRC, 2004) in extra-early maturity group for various rainfed agro-systems of the country. A large number of hybrids are in pipe line which have significantly higher yield than the existing ones.

Specialty corn: Research work on popcorn and sweet corn was started during sixties and seventies with a view to making maize cultivation more remunerative. Inbred lines together with recovered lines, derived by crossing US lines with flint lines were utilized for developing high yielding varieties. As a result, several popcorn and sweet corn hybrids/synthetics were developed. Among these, Sweet Corn Synthetic VL 15 was found to be most promising. Later in 1965, work on the development of composite strains was undertaken. A culture developed by using modified mass selection method in the original population of Amber Pop Corn procured from Hyderabad was released for UP hills under the name VL Amber Pop Corn (SVRC, 1981). A prolific early composite VL Babycorn 1 (VL 78) was found suitable as baby corn and released by CVRC for the first time for cultivation across the country.

Release of Maize Hybrids and Composites

Fourteen hybrids and eight composites have been released so far by CVRC and SVRC since 1962. Among them 'VL Makka 42' was the first maize hybrid of the country developed and released by this institute. Vivekananda Parvatiya Krishi Anusandhan Sansthan has also developed 'Vivek maize Hybrid 23' for organic cultivation for Uttarakhand hills. Among speciality corn, the institute has developed VL Amber Pop Corn with very high popping percentage as well as baby corn hybrid (VL Makka 42) and composite (VL Baby Corn 1) which are gaining popularity in South India as well as north-east due its extra-early maturity. In brief, VPKAS has released 14 Hybrids: (10 Single Cross, 3 Double Cross, and 1 Triple Cross:) and 8 Composites. Four Specialty Corn, *viz.* Pop corn: VL Amber Pop corn; Baby Corn: VL Makka 42 (hybrid) and VL Baby Corn 1; QPM hybrid: Vivek QPM 9 (Tables 4.1 and 4.2) for the mandate area as well as beyond the mandate area in different zones.

Table 4.1. Released varieties/hybrids of maize for mandate area and other parts of the country

	Recommended hybrids/ composites	
Uttarakhand	**Uttarakhand, Himachal Pradesh, Jammu & Kashmir**	**Beyond mandate area**
Vivek Maize Hybrid 9, Vivek QPM 9, Vivek Maize Hybrid 15, Vivek Maize Hybrid 21, Vivek Maize Hybrid 23, Vivek Maize Hybrid 25, VL Baby Corn 1, VL *Sankul Makka* 11, VL *Sankul Makka* 31	Vivek Maize Hybrid 9, Vivek Maize Hybrid 15, Vivek Maize Hybrid 21, Vivek Maize Hybrid 25, VL Babycorn 1	Vivek Maize Hybrid 9 (Z4), Vivek Maize Hybrid 17 (Z3, Z4, Z5), Vivek Maize Hybrid 21 (Z3, Z4), Vivek Maize Hybrid 27 (Z3, Z4), VL Babycorn 1 (All India)

Germplasm and its Exchange with Other Centres

Attempts made to improve the gene pools/populations/germplasm received from various sources have led to improvement in Pool 23, Pool 17, Pool 18, Pool 29, Pool 31 C4 HS bulk, Pool 33, Pool 40 and Pool 32 for earliness, disease resistance and other agronomic traits. These improved versions act as superior base populations for composite and hybrid development programme.

Table 4.2. Details of 22 Varieties (14 hybrids and 8 composites) of maize from VPKAS, Almora

S. No.	Varieties	Year of release	Area of adaptation	Pedigree
1.	VL Makka 54	1962 CVRC	Himalayan Hills	(CM 107/CM 108)/ (CM 203/CM 204)
2.	VL Makka 42	1972 SVRC 1988 CVRC	Himalayan Hills & Peninsular India	(V 558/V 57)/(V 208/ V 50)
3.	VL Makka 1	1973 SVRC	U.P. Hills	Compositing of genotypes
4.	Him 128	1979 CVRC	Hills of UP & HP	(CM 301/CM 401)/ (CM 302/CM 303)
5.	VL Makka 16	1980 SVRC	U.P. Hills	Compositing of 25 genotypes
6.	VL Amber	1981 SVRC	U.P. Hills	Compositing of 6 Popcorn inbreds
7.	VL Makka 41	1986 SVRC	U.P. Hills	Compositing of 12 progenies
8.	VL Makka 88	1988 CVRC	Himalayan Hills	Compositing of 15 genotypes
9.	Him 129	1997 CVRC	Himalayan Hills, Guj, M.P. & Raj	(CM 128/CM 129)/ CM 502
10.	V Hybrid 4	1999 CVRC	Guj, M.P. & Raj (RF conditions)	CM 212/CM 141
11.	V Hybrid 5	1999 SVRC	U.P. Hills	CM 212/V 25
12.	V Hybrid 9	2000 CVRC	Himalayan region, A.P., Maharashtra, T. N. and Karnataka	V 270/CM 212
13.	V Sankul M 11	2003 SVRC	U.A. Hills	Composite of 6 inbred lines
14.	VL Baby Corn 1	2004 CVRC	All India	[(VL 16 × Murulia) × VL 16) × VL 16)]
15.	V Maize Hybrid 15	2004 CVRC	Rainfed conditions of Zone 1 (Himalayan Region)	CM 152 × CM 212
16.	V Maize Hybrid 17	2004 CVRC	Zone 3 (North UP and NE region), Zone 4 (Peninsular region), Zone 5 (Western & Central region)	CM 153 × CM 212

Table. 4.2. *Contd.*

S. No.	Varieties	Year of release	Area of adaptation	Pedigree
17.	V Maize Hybrid 21	2007 CVRC	Zone I, Zone II and Zone IV (Himalayan region, A.P., Maharashtra, T. N. and Karnataka)	CM 212 × V 341
18.	V Maize Hybrid 23	2007 SVRC	Organic conditions in hills of Uttarakhand	V 351 × V 341
19.	V Maize Hybrid 25	2007 CVRC	Zone I (UA, HP, J&K and NEH)	V 341 × V 346
20.	V Maize Hybrid 27	2007 CVRC	Zone III (North UP and NE region), & IV (Mah, AP, Karnataka & TN)	V 335 × V 345
21.	V Sankul Makka 31	2007 SVRC	Uttarakhand Hills	Composite of 7 experimental & released varieties
22.	Vivek QPM 9	2007 SVRC	Uttarakhand Hills	VQL 1 × VQL 2

For harnessing the genetic gains in hybrid development, two broad genetic base pools, *viz.*, Early Heterotic Pools I and II, have been synthesized after studying the heterotic pattern of a wide array of diverse germplasm. These pools are expected to serve as superior germplasm for a long period to breed new hybrids as well as composites for the hills.

For maintaining constant flow of desirable genes to develop superior varieties, about 150 genetic accessions from U.P. hills were collected and evaluated. By grouping these collections based on their maturity, very-early, early and medium duration pools were constituted. This programme has led to the development of a very-early (80–85 days) composite VL Makka 41, which was released by SVRC in 1987. Studies revealed that native germplasm, possessing wide range of genetic diversity for economically desirable characters can be exploited for improving yield, grain yield/plant and number of grains/row.

VPKAS had developed a large number of inbred lines and heterotic pools (Tables 4.3 & 4.4) and shared (Table 4.5) them with maize group of the country. The inbreds are supplied (Table 4.6) to NBPGR, NDLS; GBPUA&T, Pantnagar; NDU Agri & Tech, Bahraich; DMR, Delhi; SKU Agr & Tech, J&K; ICAR Res Complex for NEH Region, Barapani; ARS Amberpet, HYB; ARS Rajenderanagar, HYB; BK V V Nadia, WB; JK Agri-Genetics, Sec and SFCI.

Table 4.3. Genetic resource developed at VPKAS Almora in the last 5 years

Maize	Accessions maintained
NEH Prolific	15
Exotic Pop Corn Lines	26
QPM CIMMYT Lines	23
VL Lines Converted	10 (VQL1, VQL2, VQL3, VQL5, VQL6, VQL8, VQL9, VQL10, VQL12, VQL13)
Recipient Inbred Lines	3 (CM145, CM212, V25)
Donar Lines	7 (CML162, CML170, CML173, CML176, CML180, CML184, CML189)
turcicum Resis. Lines	4 (V53, CM145, V335, V336, V356)
Good CA lines	3 (CM212, V341, CM145)

Table 4.4. Maize inbreds and populations developed as new genetic resources

Maize pools/population	Accession
CIMMYT	CM141, CM145, CM502, U15-1, Syn-I, Syn II, V6, V12, V13, V17, V21, V24, CM 152 CM 153, V178, V190, V198, V273, V322, V327, V328, V332, V333, V348, V353, V354, V359, V360, V365, V366, V367, V368, V370, V371
Exotic lines	CM212, V25, V33A, V34A, V46, V53, V281, V284, V324, V326, V329, V334, V335, V336, V337, V339, V341, V343, V347, V349, V364, V369
Elite indigenous material & promising Lines	V26, V27, V28, V49, V128, V261, V338, V340, V345, V346, V350, V351, V352, V355, V356, V357, V358, V361, V362, V363, V372, V373, V374
Hill early yellow pool (HEYPOOL)-National pool 5	Developed by VPKAS Almora
Early yellow heterotic pool I (EYHP-I)	Developed by VPKAS Almora

Table 4.5. Inbred supplied to other maize breeding centers (1997–2006)

NBPGR, NDLS	CM128, CM129, CM502CM212, CM141, V 25
GBPUA&T, Pantnagar	CM128, CM129, CM212, CM141, CM145, V 25
NDU Agri & Tech, Bahraich	CM128, CM129, CM212, CM141, CM145, V 25
DMR, Delhi	CM126, CM127, CM128, CM129, CM212, CM141, CM145, V 25
SKU Agr & Tech, J&K	CM128, CM129, CM502, CM212, CM141, CM145
ICAR Res Complex for NEH Region, Barapani	CM128, CM129, CM502, CM212, CM145, V 25
ARS Amberpet, HYB	CM126, CM127, CM129
ARS Rajenderanagar, HYB	CM126, CM129, CM212, CM141
BK V V Nadia, WB	CM126, CM127, CM128, CM129
JK Agri-Genetics, Sec	CM126, CM127, CM128, CM129, CM 502

Table 4.6. Maize frontline demonstrations *kharif* 1999 to *kharif* 2006

Year	Number of FLDs	Number of varieties/ hybrids used	Area covered (ha)	Increase (%)	
				Hybrids	Composite
1999	62	4	7.76	99–100	37–45
2000	94	4	7.80	76–141	54
2001	64	4	8.60	125–178	107
2002	189	6	12.64	51–102	41–66
2003	364	7	15.87	36–116	21–60
2004	114	6	11.20	129–227	104–131
2005	221	6	17.13	113–117	102–119
2006	364	5	14.47	83–148	107–119
Total	**1472**	**14**	**95.47**	**36–227**	**21–131**

Popularization of Maize Hybrids and Composites in Uttarakhand Hills and Other Parts of the Country

A large number of off-farm demonstrations were organized in the hills to popularize cultivation of improved maize varieties. A total of 1472 FLDs covering more than 95 hectare area in hills was conducted. The yield gain in FLDs was 36 to 227% in hybrids and 21 to 131% in composites. The details of maize FLDs are given in Table 4.5 since *kharif* 1999 leading to greater acceptability by the farmers for seed as well as fodder. Overall results of the FLDs conducted since 1999 showed conclusive evidence that all the varieties performed well in farmers' fields. Among the hybrids, Vivek Maize Hybrid 9 (3314 kg/ha) gave maximum yield advantage (125.83%) followed by Vivek Hybrid 4 (111.2%), Him 129 (86.43%) and Vivek Hybrid 5 (72.04%). Among the composites, Vivek Sankul Makka 11 showed 75.03% yield advantage, which was much more than the yield advantage obtained in VL Makka 16 (36.29%). The production of inbred lines has increased by 23.5 times from 55 kg in 1999 to 1350 kg in 2006.

Subsequently, in the last 5 years the demand for maize seed has increased for composites/hybrids (Fig 4.1) as well as production of inbreds (Fig 4.2) especially with the release of new hybrids and baby corn composites. The inbred breeder seed production and supply (Fig 4.3) of the released hybrids has also increased. The inbreds have been supplied to a number of organizations. The demand of maize seed has also increased in Assam, Hyderabad, Bihar etc. in maize single-cross hybrids as well as specialty corn.

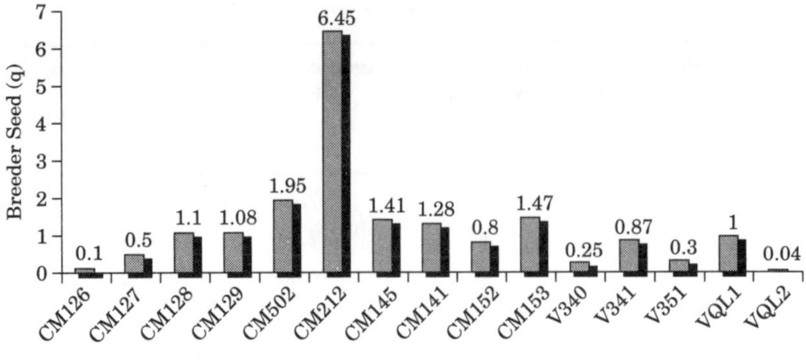

Figure 4.1. Inbred breeder seed production and supply (1997–2007).

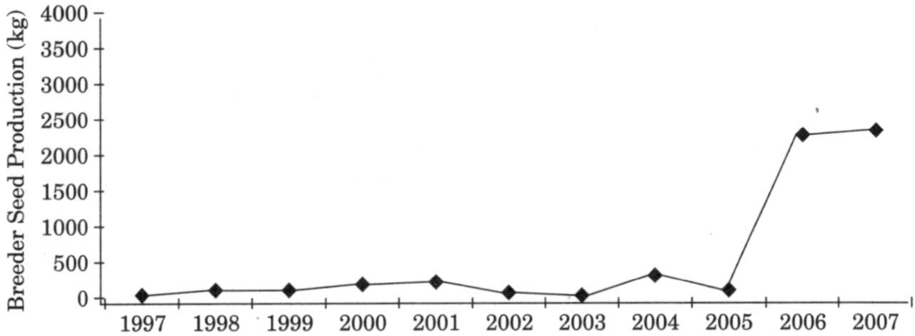

Figure 4.2. Inbred production (breeder seed) (total 9717 kg)

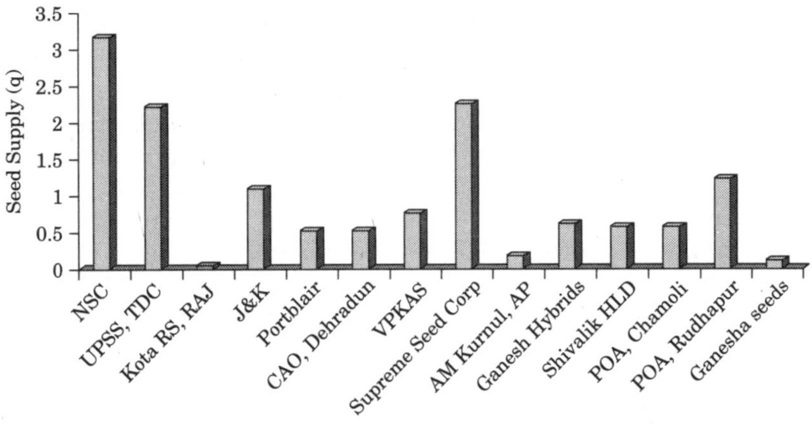

Figure 4.3. Inbred breeder seed supply (1997–2006)

Economizing Seed Production of Single Cross Hybrids by using Narrow-Based Seed Parents

Single cross hybrid though has high yield suffers from poor seed production. A scheme to develop and use narrow-based seed parent for the development of single-cross hybrid was formulated (Fig 4.4) which helped in the development of 'Vivek Maize Hybrid 15' (Zone 1) and 'Vivek Maize Hybrid 17' (Zones 3, 4, & 5) and released by CVRC in 2004. In the parentage of 'Vivek Maize Hybrid 15' and 'Vivek Maize Hybrid 17', CM152 and CM153, respectively, are narrow based female seed parents with 60-70% higher seed yield than the best hybrids available.

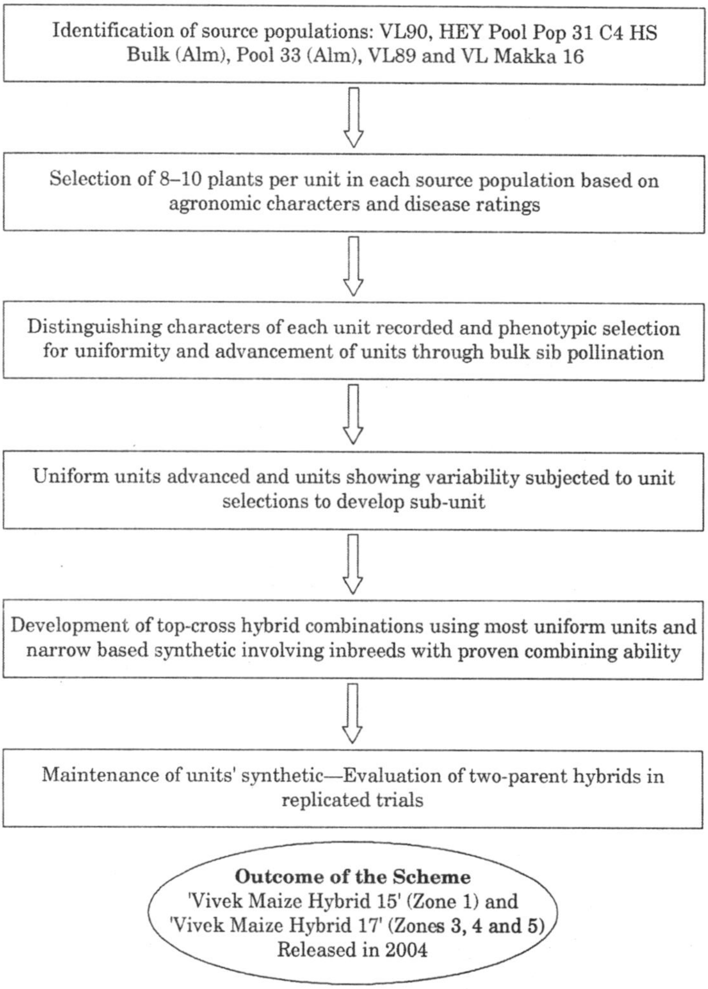

Figure 4.4. Scheme for development of narrow-base seed parents and two-parent hybrid combinations

Quality Protein Maize (QPM) for Nutritional Security

Maize plays an important role in human and animal nutrition. However, the normal maize protein is of poor quality owing to deficiency in two essential amino acids, lysine and tryptophan, high leucine-isoleucine ratio and low biological value. The *opaque2* maize did not become popular with farmers as well as consumers mainly because of reduced grain yield, chalky and dull kernel appearance and susceptibility to ear rots and stored grain pests.

Rapid advances in genome research and molecular technology have led to the use of DNA-MAS (marker aided selection), which holds promise in enhancing selection efficiency and expediting the process of varietal development. An innovative combination based on SSR markers for *opaque2* allele and phenotypic selection for kernel modification has been developed at the Plant Molecular Biology Laboratory, VPKAS, Almora, for rapid, efficient and cost effective conversion of normal single cross maize hybrids (Vivek-5 and Vivek-9) into QPM. The objective is to convert normal elite maize inbreds, which are parents of single cross hybrids into QPM through marker aided selection. Microsatellite markers, *viz., Phi 057, phi 112* and *umc 1066,* which figure as internal repetitive sequences within the *opaque 2* gene are utilized in the PCR based assays to select individuals carrying a copy of the *opaque 2* gene in successive back-cross generations. Three normal inbreds, viz., CM212, CM145 and V25 have been converted to QPM versions with tryptophan levels in the endosperm above 0.8 per cent. The single cross QPM hybrids developed using converted QPM versions of V25, CM 212 and CM145 showed 9–12 per cent superiority in grain yield over their normal versions with tryptophan concentration of above 0.72 per cent in the endosperm protein. The integrated two generation marker based back cross breeding strategy developed by Plant Molecular Biology Laboratory of VPKAS brings together the salient features of both marker aided and phenotypic selection approaches such as fixing the large segregating generation for target, reduction of the linkage drag by selecting for flanking markers of recipient allele type, recovery of maximum amount of RPG within two BC generations and provides ample scope for exercising phenotypic selection for as many desirable agronomic and biochemical traits as possible.

Table 4.6. Tryptophan content in normal and QPM hybrids developed through MAS

Hybrid	Tryptophan (%)	Tryptophan gain (%)
FQH 4567	0.83	40.7
Vivek 9 (Check)	0.59	
FQH 38	0.85	73.5
Vivek 21 (Check)	0.49	

Progenies from BC_2S_2 generation of a cross between CM 145 (Non QPM) and CML 176 (QPM). 25% opaque endosperm lines contain elevated levels of tryptophan and lysine besides normal appearing kernel.

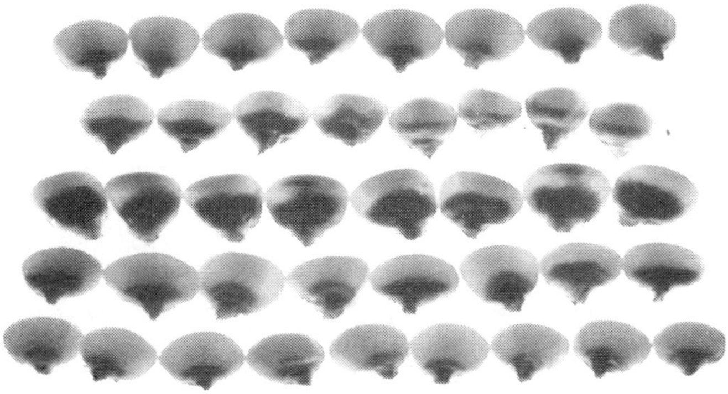

Outcome

'Vivek QPM 1 (Uttarakhand Hills)
(FQH 4567) Released in 2007

Marker Aided Selection for Quality Protein Maize (QPM)

QPM is a genotype in which *opaque-2* gene has been incorporated along with its associated modifiers and contains twice as much lysine and tryptophan than normal maize. Three SSR markers, *viz., phi 057, phi 112* and *umc 1066* are utilized in PCR based assays to select *opaque-2* gene responsible for high lysine and tryptophan.

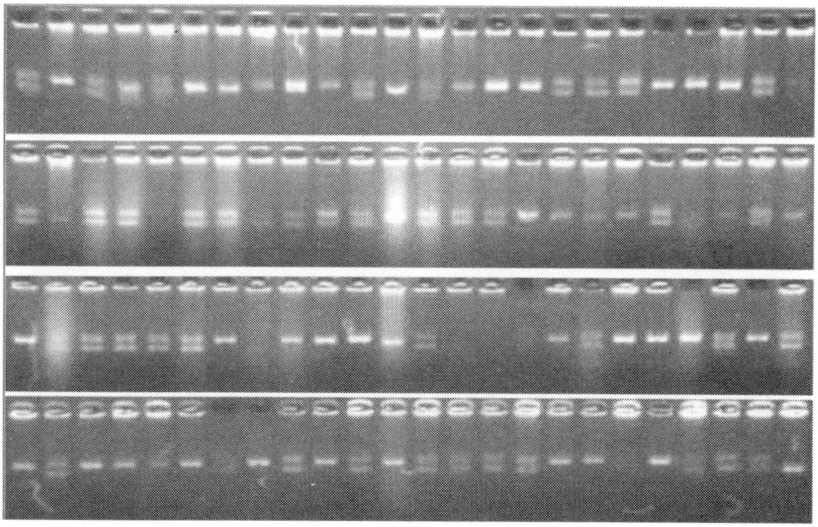

Genotyping of back cross (BC_1) progenies of QPM X non QPM crosses using *umc 1066.* Double bands indicate desirable heterozygotes and the single bands indicate discarded homozygotes

Phenotypic Selection for Endosperm Modification

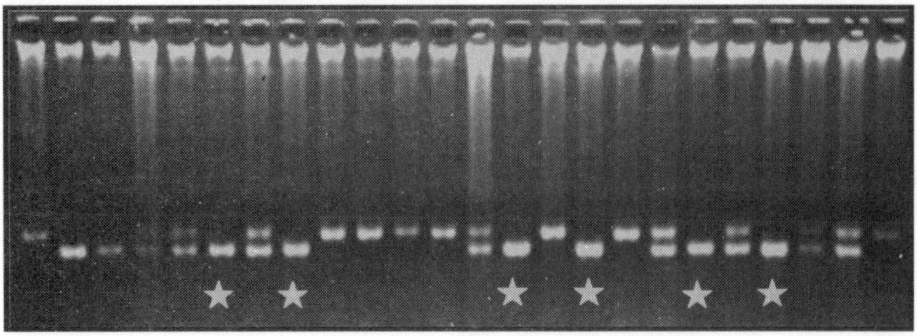

Identification of homozygous recessive individuals in BC_3S_2 generation. The individuals marked by★ are homozygotes for recessive *opaque-2* allele.

MOLECULAR BREEDING SCHEME FOR QPM (BABU *ET AL.*, TAG, 2005)

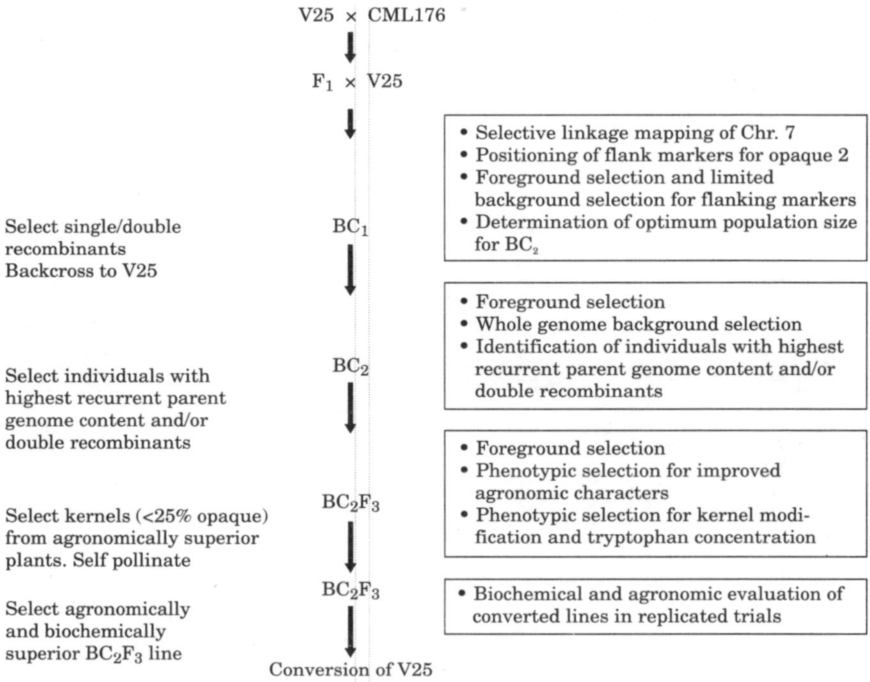

Strategies for the Commercialization in Maize

A multiple approach needs to be followed to increase the commercialization of the hybrids that will facilitate the availability of superior hybrids to farmers.

Release of more number of hybrids: More number of hybrids in different maturity groups may be released belonging to different organizations in different agro-climatic zones. Instead of one hybrid per zone more than one hybrid belonging to different organization may be released. This will stimulate competition among different organizations to provide superior hybrids to the farmers.

Economically profitable hybrid seed production: The availability of hybrid seed will enhance only when the farmer himself is involved in the hybrid seed production and his hybrid seed could sell in at higher market price. The seed price should be fixed accordingly.

Training the farmers for maize hybrid seed production: A mission-oriented on-farm training programme needs to be followed to educate farmers not only for hybrid seed production in maize and other crops but also on norms and organizations dealing in seed production. VPKAS has taken a lead in this direction to produce hybrid seed at farmers field in Uttarakhand hills. In *kharif* 2007, the farmers of Liti and Sama villages in Uttarakhand hills produced 4.5 q seed of two maize hybrids (Vivek Maize Hybrid 15 and Vivek Maize Hybrid 21) under the guidance of scientists of VPKAS, Almora.

Less stringent sharing norms: Extensive sharing of promising inbred parents of popular single-cross hybrids among public and private organizations needs to be promoted by developing mutual confidence between the two organizations.

Interaction meets: Regular meets, at least twice a year, for commercialization workshops/field meeting need to be organized wherein competent authorities from different organizations should be able to take decisions in an open discussion regarding commercialization of single cross hybrids and sharing of inbreds.

5 Strategies for Commercialisation of Indian Sorghum Hybrids Bred in Public Sector

N. Seetharama[1], Vilas A. Tonapi[1], B. Dayakar Rao[1], S. Ravikumar[1], Ravinder Reddy, Ch.[2] and M. Elangovan[1]

ABSTRACT

Shift from coarse grain to fine grain is a welcome change, the production-demand equation of food grains in the country does not favour the importance of sorghum as food grain. Generation of adequate demand for kharif sorghum, cultivar development with specific grain attributes, high productivity to enhance cost competitiveness, mission oriented R and D activities involving client institutions, state of the art know how and appropriate governmental policies for promotion of production, ways to improve the compatibility of kharif sorghum by creating demand for grain and substantially bridging the gap between cultivar production potential and farm yield with the aid of cost-effective production technology and value-addition to the crop with sweet stalk, alternate areas for successful seed production programme, development of potential hybrid, retention of area under kharif sorghum, solution to the grain molds through biotech tools and strengthen the rabi hybrid development programme are discussed to regain the glory of sorghum.

Key words: Hybrid sorghum, kharif sorghum, male sterility, seed system, value addition

---

The power of a seed is unlimited. As a powerful agent of change, seeds can be a means of overcoming production constraints, thereby making a difference in the lives of the poor and hungry. This requires seed demand and supply to be balanced by way of a secure seed supply system. This would give farmers access to adequate quantities of good quality seed of the desired type at the required time and at affordable cost. Seeds are key components in the conservation and ownership of biodiversity. Accordingly, sustainable seed supply and implementation of seed security are among

1. Directorate of Sorghum Research, Rajendranagar, Hyderabad 500 030.
2. International Crops Research Institute for Semi-Arid Tropics, ICRISAT, Patancheru, A.P. - 502 234.

the major activities outlined in the Global Plan of Action for the Conservation and Sustainable Utilization of Plant Genetic Resources for Food and Agriculture (Global Plan of Action 1996). And seed therefore represent hope for the future of mankind. Throughout our history, farmers' informal seed systems have had a great influence on the evolution of modern agriculture, by practising conservation of agro-biodiversity at the gene, farmer and ecosystem levels. Therefore, investing in a range of approaches in order to strengthen seed delivery models and uptake of public bread hybrids assumes great urgency. While the formal hybrid seed industry led by the private sector has tended to focus on profit-making species and crops, the informal sector has concentrated on crops–mainly self- or open-pollinated varieties–that are crucial to local food production systems. Given such a scenario, a national seed policy concludes helping to strengthen the informal sector. International support too continues to be mainly engaged with the formal sector. Perhaps matching support is required to encourage continued development of informal seed systems.

An effective seed supply system is necessary to make good quality seed available to farmers at the right time and at low cost. Given the critical role played by improved varieties in increasing production in conventional cropping systems, decision-makers in agriculture have the challenge of developing an integrated and cost-effective seed system that is capable of generating and delivering improved seed varieties to farmers. Such a system would be an important step toward ensuring seed security and enhancing livelihoods, particularly of dryland farmers. With the concerted efforts of agricultural scientists, planners and farmers, the total Agricultural production level has crossed 200 million tonnes in India.

The area under sorghum has declined in every continent except in Africa and Mexico, while global production and productivity declined in all countries except China, Burkina Faso and India. The increase in sorghum productivity in China (59%) compared with that of India (15%) during 1981 to 2005 is a notable development. The yield gaps in India are higher due to varied soil, climatic conditions across the various zones it is being cultivated.

At the same time, in spite of the implementation of many population control schemes, human population in India has crossed the one billion mark and is projected to reach 1.3 billion by 2030. The nation will need nearly 260–264 million tonnes of food grain, excluding the requirements for seed and export, which is a formidable task for agricultural scientists to achieve. In order to meet the projected demands, average yield of food grains will need to be increased by about 1 tonne per ha between now and 2030. This gap can be bridged by: (i) increasing the land under cultivation and (ii) increasing the yield per ha. The first option is not feasible due to increased demand for land under roads, houses and factories. Moreover, the arable earth surface is deteriorating and decreasing as a result of soil erosion, salinization and acidification, all resulting in reduction of land available for cultivation from the present 170 million ha to 100 million ha by 2030.

It is possible to increase the gross sown area, to some extent, through increasing the cropping intensity. The second option may be achieved and is possible by: (i) ensuring availability and efficient use of water, fertiliser and plant protection measures, (ii) timely planting of quality seeds and ensuring desired plant population, and (iii) development of more productive varieties. When we talk about quality seed, it includes two components: (i) seed of improved variety and (ii) seed quality parameters.

International Scenario

India is the largest sorghum grower in the world (area of 9.33 million ha 2005–06) followed by Nigeria and Sudan (Fig 5.1). In case of sorghum production, the situation is reverse, India ranked seventh among the nine countries growing more than 1 million ha under sorghum. During the 2005–06, China topped in productivity with 3960 kg/ha. Indian sorghum productivity during the same period was 20% of the productivity level of China or 60% of the global productivity (Table 5.1).

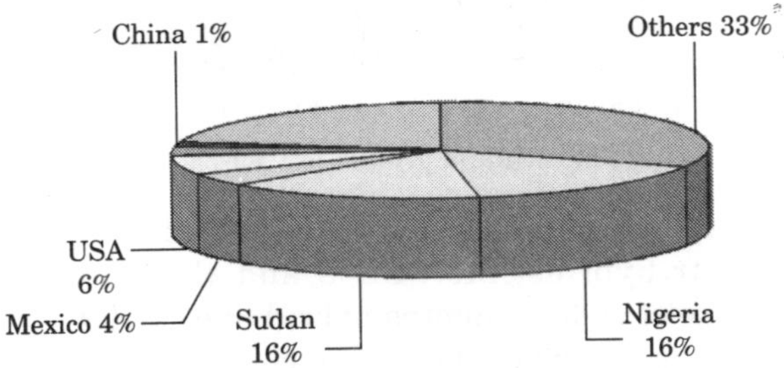

Figure 5.1. Distribution of sorghum growing area across the world (2005–06)

Table 5.1. Comparison of Indian sorghum production with the rest of the world

Countries	Area (000 ha)			Production (000 t)			Yield (kg/ha)		
	1981	2005	% #	1981	2005	% #	1981	2005	% #
World	44.89	44.37	–1.16	65.52	58.99	–9.97	1458	1330	–8.779
Nigeria	2.68	7.07	163.51	3.28	8.03	144.52	1461	1135	–22.31
Sudan	3.05	7.03	130.19	2.27	4.01	76.42	725	565	–22.06
Mexico	1.49	1.9	27.43	4.99	6.59	32.04	3343	3471	3.82
USA	5.27	2.71	–48.61	19.16	10.67	–44.30	3618	4000	10.55
China	2.83	0.66	–76.66	7.03	2.6	–63.04	2493	3975	59.44
India	16.36	9.33	–42.97	11.38	7.47	–34.36	695	801	15.25
B. Faso	1.05	1.52	44.76	0.62	1.47	137.10	589	969	64.51

Change in 2005 over 1981

Source: FAO database

National Scenario

The sorghum area in the country, which was 15.5 million ha in the early 1950's, increased to 18 million ha during the same decade and almost remained unchanged throughout the 1960's. During this period, the total production rose from 5.7 million tons to 9.8 million tons due to continuous rise in the productivity from 370 kg to 530 kg/ha. Since then there was a continuous drop in the area. Almost 1 million ha area was lost between 1970 and 1973 and another 1 million ha between 1988 and 1989. During the last 10 years, the reduction in area continued and the current area is 9.2 million ha. Therefore, during the last 15 years, a total area of 6.9 million ha has been diverted to other crops (Fig 5.2).

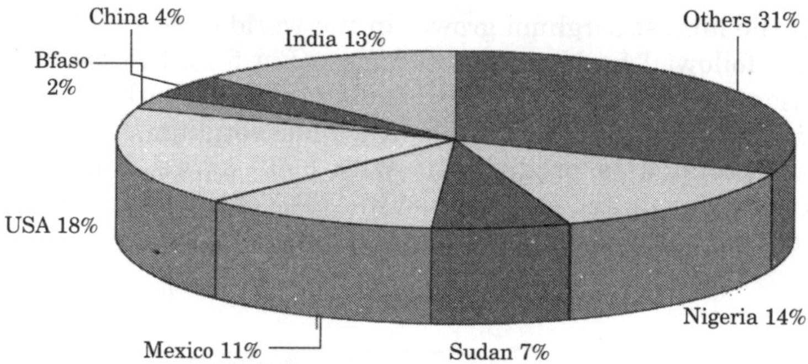

Figure 5.2. Share of sorghum production across the world (2005–06)

The first set of hybrids was released in the mid-1960's. However, the release of popular hybrids like CSH 5 and CSH 6 in mid-1970's and subsequent release of widely adopted hybrid like CSH 9 in the early 1980's have augmented the spread of HYV's and consequent rise in productivity. This could have made a major impact in the production although it was constantly being neutralized with the negative growth rate of area. As a result, the total production of 9.67 million tons in 1975–76 increased to only about 12 million tons in favorable years. The area and production during the 2005 are 9.20 million ha and 7.47 million tons, respectively with 816 kg/ha productivity. The season-wise and state-wise sorghum production trends and sorghum scenario are depicted in Figs 5.3 and 5.4, respectively.

State-wise area, production and yield of sorghum during 2005 are given in Figs 5.3 and 5.4 and 5.5. Currently, Maharashtra is the largest sorghum producer with 1.51 million ha area under kharif and 3.10 million ha under rabi. In Karnataka, out of 1.63 million ha of total sorghum area, 1.31 million ha is under rabi and the rest under kharif. Sorghum area in Madhya Pradesh is 0.60 million ha and in Andhra Pradesh it is around 0.48 million ha. The productivity of Karnataka (1345 kg/ha), AP (1336 kg/ha) and Maharashtra (1254 kg/ha) and is the highest during kharif, while the rabi productivity is high (1153 kg/ha) in Andhra Pradesh.

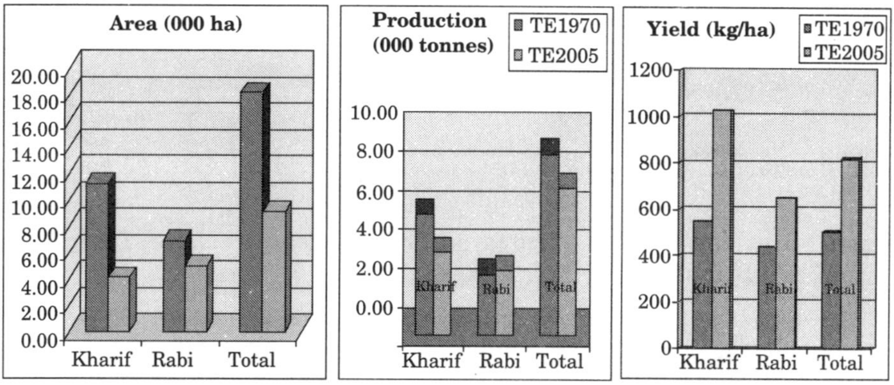

Figure 5.3. Trend in sorghum area, production and yield in India

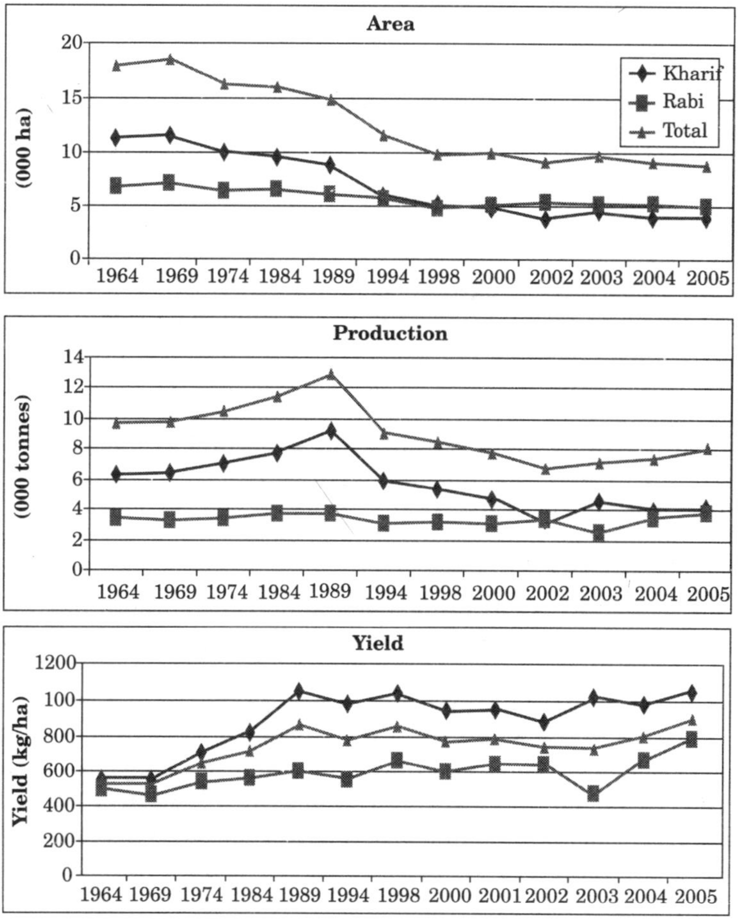

Figure 5.4. Season-wise changes in area, production and productivity of sorghum in India

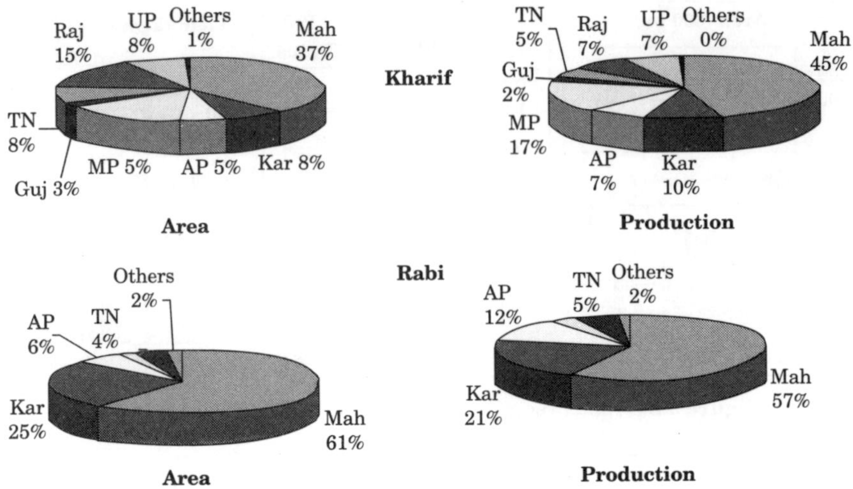

Figure 5.5. Area and production of sorghum in major sorghum growing states (2005)

The success story of Maharashtra during kharif suggests that lack of HYVs per se is not the primary reason for the low national productivity. While certain agronomic inadequacies of kharif HYVs and gaps in the development of rabi HYVs could be pointed out as reasons for their regional non-preference, the yield gaps existing between their production potential, farm demonstration yield and average farm yield have other reasons. Some of the most visible reasons are the limitations imposed by national resource factors; biological limitations; governmental policy; and static research strategy.

Achievements in the Development of Sorghum Hybrids

Kharif sorghum: Varietal improvement leading to the development of high yielding hybrids and varieties forms the central theme of research achievements.

With the release of CSH 1, the first commercial hybrid in 1964, sorghum became the second crop after maize in developing high yielding hybrids using cytoplasmic-genetic male sterility system. Since CSH 1, nineteen more hybrids were centrally released. A few more hybrids adapted to specific regions were released at state levels. Hybrids CSH 1 to CSH 23, are a standing testimony of success of Indian sorghum breeding not only in terms of yield enhancement, but also in terms of diversification of parental lines and progressive advances in the incorporation of resistance against major pests and diseases. The hybrids played a major role in pushing up productivity and production, particularly in the case of kharif sorghum. Among the kharif hybrids, the role played and being still played by CSH 1, CSH 5, CSH 6, CSH 9, CSH 14 and CSH 16 needs special mention. While

CSH 5 and CSH 6 had a yield potential of 34 q/ha, this potential was raised to 40 q/ha in CSH 9. It is now further advanced to more than 41.0 q/ha in CSH 16 and CSH 23.

Introduction of hybrids witnessed a major change in Indian sorghum farming specifically in kharif season, which was traditionally saturated with land races. These land races were tall (250–280 cm), low yielding, with long duration (130–140 days), input non-responsive and with low harvest index (15–17%). The hybrids, on the contrary, were dwarf (170–180 cm), vigorously growing, input-responsive, shorter in maturity period (100–110 days) and very high yielding with high harvest index. They brought in physiological efficiency in the uptake and utilisation of nutrients, increased sink size and efficient partition of dry matter between panicle and stalk.

CSH 1 was developed from exotic parental lines with grain and plant attributes not preferred in India. As many of the critical production traits were not available in Indian land races, but available only in exotic temperate germplasm, a major intercross of these material were undertaken to generate new variability to suit to the Indian needs. While these initial crosses used the material introduced from the United States of America, the later broad based diversification of breeding material largely depended on the germplasm belonging to the Ethiopian *Zera zera* type. This breeding programme with primary emphasis on kharif sorghum led to the land mark successes in the development of ms lines such as 296 A, 2077 A and 2219 A and R-line such as CS 3541. Moving away from CSH 1, the hybrids not only gained in yield, but also in their tolerance to major pests and diseases. All hybrids from CSH 5 onward were bred with simply inherited tan plant character which is associated with resistance to foliar diseases. Hard grain in this plant background helped to improve resistance to grain mold.

The dwarf hybrids, in general, are poor fodder yielder in comparison with the land races. As fodder is an economically important component, development of dual purpose hybrids and varieties with high fodder and grain yield also received attention. CSH 10 and CSH 13 are the hybrids of this nature.

The shift in maturity of the high yielding hybrids and varieties had important implications in ensuring better yield. While the traditional land races could be classified under late types (130 or more days), the HYVs fell in medium (110 days) and early (100 day) maturing types. Among the popular hybrids CSH 1, CSH 6 and CSH 14 are early types and CSH 5, CSH 9, CSH 13, CSH 16, CSH 18 and CSH 23 are medium types. These maturity groups helped in their adaptability to the growing conditions differing in soil moisture retention, rainfall and cropping system. Reduction in maturity helped to avoid terminal moisture stress associated with low rainfall and shallow soil profile. Heterozygosity of hybrids is also considered to have conferred better adaptability under conditions of stress.

Breeding also let to the development of high yielding varieties (CSV 1 to CSV 18), many of them with medium maturity, for the Central release. Many more varieties were also released in various States. Some of these varieties are dual purpose type. By and large, varieties encountered less acceptability among farmers. Better preference was received by dual-purpose varieties such as CSV 10, CSV 13, SPV 462 and CSV 15 in some restricted pockets. A major advantage of varieties over hybrids was their relative better grain quality and multiple resistance or tolerance against major pests and diseases. The recently released varieties, CSV 15 to CSV 18 could establish higher grain and fodder yield potential than the hybrids CSH 5 and CSH 6 released two decades back. However, a short duration variety CSV 17 has been released from the Udaipur centre.

Rabi sorghum: Improvement of rabi sorghum until the nineties did not receive as much emphasis and effort as the kharif sorghum. However, some of the hybrids and varieties listed in Appendix IV are specifically developed and recommended for rabi season. Five hybrids and five varieties were hitherto centrally released. In rabi sorghum, the fodder yield has more importance than that in kharif sorghum. Therefore, rabi grain productivity has to be accompanied by normal or better fodder productivity. From this point of view, a progressive success was achieved from the first rabi hybrid CSH 7R to the latest hybrid CSH 19R. Unlike the kharif cultivars, higher levels of resistance against major pest (shoot fly) and disease (charcoal rot), stringent maturity duration to suit different receding soil moisture regimes and certain levels of thermo-insensitivity are essential in rabi cultivars for better adaptability. Grain quality is also as much important as the grain yield. The quality bench mark is that of the popular land race, Maldandi 35-1.

In adaptability criteria such as shoot fly resistance as well as the grain quality aspects, the varieties are superior to hybrids. Unlike kharif, it appears rabi varieties have better preference over rabi hybrids for reasons of adaptability and grain quality. The four rabi varieties released, CSV 8R, CSV 14R, CSV 18 and Swati, were better received than the rabi hybrids such as CSH 7R and CSH 8R. Recently developed hybrids CSH 15R and CSH 19R are more productive and adaptive with higher acceptability among farmers.

Forage sorghum: Earlier evaluation of germplasm accessions of about two thousand lines revealed that *Sorghum roxburghii* types possess desirable traits including resistance to leaf spot diseases, low tannin content and good combining ability for quality traits. Early forage sorghum varieties like Haryana Chari, SC 136 (single-cut) and Meethi Sudan SSG 59-3 (multi-cut) were developed at the Hisar Centre. The multi-cut variety, Meethi Sudan with a potential of four cuts became popular due to its high green foliage, yield potential, regeneration and excellent forage attributes.

Subsequently, another ten varieties namely MP Chari, UP Chari 1, UP Chari 2, RC 1, RC 2, PC 9, PC 23, HC 171, HC 260 and HC 308 were developed for all India cultivation and few other varieties released at state level provided the base for varietal transformation in forage sorghum. The latest varieties are improved in terms of resistance to leaf spot diseases and stem borer and seed yield. Seed yield improved up to 14 to 15 q/ha under north Indian conditions. These varieties also exhibited higher per day productivity and improved *in vitro* dry matter digestibility (IVDMD), dry matter digestibility (DMD) and total soluble sugars (TSS) with comparable protein content. Release of dual purpose sweet (TSS 16%) and stay green variety CSV 15 for green forage production in north-west India is expected to ensure better seed supply and nutrition to livestock. Development of multi-cut forage sorghum hybrid is still to create impact on forage production, though intensive work is underway at few centres. Recent multi-cut releases include PC 106 for all India and Punjab Sudan for Punjab, in addition to the latest central release of Pant Chari 5 (UPFS 32). Recently, during 2005 a multi-cut forage hybrid CSH 20MF has been released at national level.

Sweet stalked sorghum: Till date the SSV 84 was the only national released sweet stalked variety. There are a few private hybrids in the market like Madura and Sugar Gaze. Realising the importance of having a high yielding superior sweet sorghum hybrid, the national programme nationally released the first sweet stalked sorghum hybrid CSH 20SS. Efforts are on for development of sweet stalked sorghums for various specific end users like production of ethanol, syrup and jaggery.

Breeding for value-addition of kharif sorghum: Apart from possessing the C4 photosynthetic system, sorghum is notable for high rate of dry matter production. Among all *Graminaceous* crops, sorghum ranks second after Napier grass in dry matter production rate. It is estimated that, on an average, sorghum can produce 59 g dry matter/day/m^2 in a number of world locations. Thus it has the physiological potential as an economically viable biomass crop. Moreover, it has ability to accumulate large quantity of sugars in the stalk. The so called sweet sorghum may produce higher biomass yield and stalk sugar content. Sweet sorghum genotypes with yield potential of 100 t fresh stalk/ha with 18–20 total soluble solids (TSS) in the stalk juice in 130–150 days is reported from elsewhere. Sweet sorghum stalk can be industrially processed for production of ethanol or jaggery or syrup and the stalk residue used either as fodder or fuel for electricity generation or to manufacture paper or particle board while the grain is used as food or feed or industrial raw material. Alcohol production from sorghum stalk and grain is more environment friendly than from molasses. This alcohol is free from the undesirable attributes of molasses based alcohol for potable use. Consequently, value addition to the crop and economic gain is far higher than that from grain sorghum.

Researches on sweet sorghum established that the stalk sugar content is a stable character when grown without moisture stress and there is enough variability to improve the biomass yield, stalk juice extractability, sugar content in juice, ratio of sucrose and reducing sugars (sugar ratio), and grain yield. It is also established that genotypes can be selectively bred with capability for high dry matter production and its efficient partitioning between stalk and panicle to achieve high fresh stalk yield, high sugar content and high grain yield. Explorative studies on heterosis in sweet sorghum hybrids found that while adequate heterosis can be generated for biomass and grain yield, the level of achievable heterosis is low for stalk sugar content. The clue to develop hybrids with high TSS is incorporation of high TSS in both parents. NRCS, for the first time developed a number of sweet sorghum based (high TSS) MS and R lines.

These researches so far led to the identification and release of two varieties. SSV 84 was centrally released and HES 4 was released in Maharashtra. Research on utilisation of sweet sorghum stalk developed methods for production of jaggery and white syrup. A crop yielding 40 t fresh stalk/ha with 18 to 19 TSS and 60% extractability would yield about 3 tons (t) jaggery and 3.5 t syrup per hectare. The same stalk yield in fermentation may yield 2500 t of ethanol. The stalk residue left by these processing is about 12–15 t/ha which may be used either as fuel or feed.

Progress of Hybrid and Varietal Research

In India, the variety improvement programme was initiated in the 1930's. The local genotypes were tall, late maturing, flowering after the rainfall seized, generally photosensitive and characterized by localized adoption and low harvest index. Their response to improved management in terms of the increased yield was very poor. Most of the improved varieties were the result of pure line selection practiced in the principal local varieties. Local x local hybridization followed by selection resulted in varieties with marginal increase in grain yield. Notable among these varieties developed during the early period and still under cultivation are the Co-series in Tamil Nadu; the PJ kharif and rabi selections, *Saoner*, *Ramkel*, *Aispur*, the *Maldandi*, *Guntur* and *Anakapalle* series of Andhra Pradesh; the *Bilichigan*, *Fulgar White*, *Fulgar Yellow, Kauvi, Nandyal, Hagari,* and *Yanigar* varieties of the erstwhile Mysore State (Ganga Prasad Rao and House, 1972).

With the discovery of a workable cytoplasmic-nuclear male-sterile system and initiation of the accelerated sorghum project (which later became All India Coordinated Sorghum Improvement Project) in 1962 hybrid breeding was given due emphasis. Initially, the germplasm from the USA (*kafir* milo cytoplasm and other germplasm) and hybrid combinations by making temperate × temperate crosses were tested. During 1962–69, out of temperate × temperate and temperate × tropical crosses, three hybrids, *i.e.* CSH 1, CSH 2 and CSH 3 were released.

Introduction of CSH 1 in farmers' fields during the 1960's resulted in quantum jump in productivity and production as the hybrid responded well to improved management practices as compared to old varieties. It became popular with farmers as it had high yield potential, suited to light soils and low rainfall areas. The second hybrid CSH 2, was based on the same male sterile, CK 60, as that of CSH 2 and a new R line IS 3691, which was a yellow endosperm selection of *Hegari* from USA. Later, a new male-sterile line, 2219 B, was identified from the germplasm line *Kafir Shallu* and a hybrid CSH 3, was developed by using 2219 A and IS 3691 (R line of CSH 2). CSH 2 and CSH 3 did not become popular due to seed production problem.

During the next decade (1970–79), three hybrids, CSH 4, CSH 5 and CSH 6 were released. Hybrid CSH 4, based on ms 1036 A had better fodder yield. The male sterile line of this hybrid, *ms* 1036 A was developed from a cross of CK 60B and PJ 8K (a local variety from Maharashtra) and the R line was a selection from IS 3924. Though this hybrid was good for grain yield, the hybrid was not popular among the farmers as its grain quality was not better than that of CSH 1.

In general, grain yield of improved cultivars was tripled by utilization of exotic breeding material in hybrid programme, however, these cultivars especially hybrids were not well received by the farmers because of the increased susceptibility of hybrids to major pests and diseases and inferior grain quality of the hybrids compared to locals.

Keeping the above factors in mind, new male-sterile lines were developed with different genetic backgrounds. The *ms* lines and R lines were developed from derivatives of crosses between temperate × tropical crosses. The *ms* line of CSH 5, (2007B) was developed from IS 2046, a germplasm line from Senegal. The hybrid CSH 5, based on converted lines *ms* 2077 A and CS 3541, contributed not only to substantial yield improvement. The hybrids were tolerant to mites and aphids also. This improvement was achieved by introduction of genes from tropical material IS 3541. An early maturing hybrid CSH 6 developed from early MS line 2219 A and CS 3541 became very popular in inter-cultivation with pigeonpea.

Further increase in grain yield was achieved by development of hybrids like CSH 9, CSH 10 and CSH 11 based on new MS line 296 A during 1980–89. Indian germplasm line *Karad local* was crossed with American material IS 3922 to develop the MS line 296 A, which was the best combiner. This MS line has a very compact panicle with more number of primary branches. CSH 9, a medium duration hybrid is widely adapted and extensively grown.

The hybrids CSH 10 and CSH 11 developed from the same *ms* line, 296 A, showed marginal superiority for grain yield.

During the 1990's, most of the hybrids tested in the all-India trials were based on 296 A with various restorers but could not make any remarkable dent for grain yield over CSH 9. Though there was no significant grain yield improvement, useful diversification for early maturity and higher fodder yield had been achieved earlier with the release of CSH 13 and CSH 14. The fodder yield of CSH 13 is 40% more than that of CSH 9 although its grain yield is marginally improved. The R line of this hybrid, RS 29 that contributes to heterosis for fodder yield is developed from SC 108, an American elite line and SPV 126 (a tall mutant of CS 3541). Another, hybrid CSH 14, is about 10 days earlier than CSH 9 while maintaining same level of grain yield. The need for diversification of female parents was felt in view of seed production problems and stagnating yield level. Another high yielding hybrid, CSH 16, was developed from new *ms* line 27 A, and R line C 43. This hybrid showed further improvement in grain mold tolerance as the new genes from Ethiopian germplasm line IS 23549 was introduced into its R line. The R line has very compact panicle. In contrast to 296 A, ms line 27 A of this hybrid produces long and loose panicle with bold and round seed likes that of post-rainy season genotype that has consumer preference. 27 B has been developed from multiple crosses using germplasm lines like IS 3687, IS 3922 and 2219B. Recently two hybrids CSH 17 and CSH 18 were released. Though, the yield levels are on par with CSH 16, these were diversified for early maturity and high fodder yield, respectively. By utilizing local variety Vidisha 60-1 at Indore centre, the ms line of CSH 18 was developed. The local variety, Vidisha 60-1 not only contributes for high stover yield but also for improved grain quality. The R line of this hybrid, Indore 12 is developed from multiple crosses of SSV 53 and SPV 475 germplasm lines (Seetharama *et al.*, 2007).

Several high yielding hybrids developed by private seed companies were also released by the CVRC through the ICAR testing system. Some of them are JKSH 22, JKSH 528, PAC 54003, SVSH 99, MLSH 151, MLSH 64, GK 4009, GK 4013, G 4245, Mahabeej 177, Kaveri 388, VJH 545, VJH 550, Shrihari 555, POC 404, POC 406 and others. In addition, hybrids released through their own testing system are produced and marketed by the private seed companies. Incidentally these are also equally popular among the farmers.

Whenever there was a change in male sterile line the yield benefit was obvious. There is a need for developing new male sterile lines having better combing ability in comparison to that of the available ms lines. So far, several germplasm lines of different botanical races have been utilized in the development of parental lines. The grain yield levels of rainy season hybrids have reached the plateau and there is a need to exploit unused germplasm to diversify the genetic base.

Simultaneously, variety improvement was achieved by introducing temperate and tropical materials. The first variety, CSV 1 is a direct introduction of the American line IS 3924. By crossing temperate and tropical germplasm, subsequent varieties CSV 2 and CSV 3 were developed. CSV 4, which was used as restorer of three most popular hybrids, CSH 5, CSH 6 and CSH 9, became a very popular variety. The variety is a converted line of the African germplasm line IS 3541, and developed by crossing it with a USA germplasm line, IS 3675. CSV 5 is another variety developed from Indian local and US line IS 3687. It records striga resistance. CSV 10, the next variety, which became popular for high fodder value, was developed from a cross between the American elite variety SC 108 and Indian elite variety CS 3541.

Another variety which became very popular, SPV 462, (from Coimbatore) developed from a multiple cross involving IS 2947 and IS 3687 from USA and IS 1151 and BP 53, locals of Maharashtra and Gujarat in India, respectively. The variety is high yielding for grain and fodder with good grain quality. CSV 13, yet another variety developed from multiple cross having exotic and local genetic base is high grain yielding with medium height. The latest, CSV 15 developed from SPV 462 and CSV 13, is a dual purpose variety having grain yield equal to that of hybrid CSH 5 and fodder yield equal to that of CSH 10.

Germplasm has been used for trait based breeding in rainy season sorghums especially for disease and pest. Though the present day varieties and restorer lines posses moderate level of resistance, the male sterile lines are highly susceptible. Genetic stocks such as IS 14387, IS 14374, and IS 25017 were extensively used for incorporation of grain mold resistance into male sterile lines. Recently, the grain mold tolerant ms line 219 was developed by crossing AKMS 14B (ms line of hybrid CSH 14) with grain mould resistant line IS 14387. Similarly, many grain mould tolerant lines, such as GMRP and SR, were developed at Parbhani and Surat centers. Important resistant lines are GMRP 9, GMRP13, SR 839, and SR 384. Similarly, shoot fly fly resistance material developed at Rahuri and Akola centers of AICSIP by using germplasm lines. Seed of the varieties bred for rainy season is not as remunerative as that of rabi varieties because the seed is not bold and lustrous. Now, efforts are being made to breed varieties with bold and lustrous seed by using *guinea caudatum* and *durra caudatum*. Some of the germplasm lines contributing bold grain size are IS 51, IS 3142, 9742, IS17600, IS 19305, IS 31690, etc.

The attempt to develop heterotic hybrids for post-rainy season whole grain quality matches with that of local variety Maldandi has met with limited success, Maldandi has a unique feature which combines good grain quality with drought and shoot fly resistance which are required for rabi season cultivars. The hybrids like CSH 7R, CSH 8R, CSH 12R based on *ms* lines developed from rainy season lines. And R lines developed from local × US lines did not show much superiority over M 35-1 for yield and quality. Another rabi hybrid, CSH 13, is superior to the rabi local for grain yield but inferior to M 35-1 in grain quality. Recently released hybrid CSH 15R, is based on rabi adapted MS line having good grain quality and shoot fly tolerance. The R line of this hybrid is developed from rabi locals and CS 3541, an elite kharif variety.

Post-rainy season varieties were developed by crossing Indian locals, M 35-1 and IS 2644, with American germplasm lines. Marginal improvement was achieved for grain yield over the most popular local variety M 35-1. The recently released variety, CSV 216R is a landrace selection from rabi germplasm from Maharashtra.

New sorghum cultivars for commercialization: The new sorghum cultivars that need to be popularized are given below:

A. Grain sorghum

Name of cultivar	Year of release	Grain yield (q/ha)	Fodder yield (q/ha)	Duration (days)	Area of adaptation	Salient features
HYBRID						
CSH 19R	2000	30	60	117	All rabi sorghum growing areas of the country, suitable to medium to deep soils	Non-tan, juicy stem, broad leaves, semi-compact panicle, pearly white, slightly flat medium size seed
CSH 23	2005	41	87	99	Kharif sorghum growing areas of AP, Gujarat, Karnataka, MP, Maharashtra, Rajasthan and UP	Tan, tall (180cm), dull green midrib, medium bold, white seed. Early maturity (103 days) avoid terminal drought
SPH 1567 (CSH 25)	2007	43	127	11	Maharashtra and all Sorghum growing regions in Kharif	Plant height 205 cm, tan, long and semicompact panicle with straw coloured glumes, seeds medium sized and pearly white
VARIETY						
CSV 216 (Phule Yashoda)	2000	21	59	123	Recommended for cultivation during rabi *viz.*, AP, Gujarat, Karnataka, parts of MP, Maharashtra, & TN	Plants non-tan type, grains bold, slightly flat and creamy. Highly fertilizer responsive under deep soils/ irrigated conditions
CSV 17*	2002	25	68	97	All kharif sorghum growing areas of the country	Early maturing, tan, dark green leaves, white dull midrib colour, moderately resistant to shoot fly and stem borer
CSV 18 (Parbhani Jyothi)	2005	35	87	110	For Zone-II Maharashtra, Karnataka, AP under irrigated conditions	Tall (227 cm), thin stem with non-senescence habit, dull midrib, non-lodging and aphid tolerant

Contd.

Name of cultivar	Year of release	Grain yield (q/ha)	Fodder yield (q/ha)	Duration (days)	Area of adaptation	Salient features
CSV 20	2006	32	133	108	Recommended for cultivation during kharif in, AP, Gujarat, Karnataka, MP, Maharashtra, TN, and UP	Tall, semi-compact panicle with dense clustering of grain in panicle branches, panicle shape oblong and pearly white bold seed. Resistant to leaf diseases
CSV 22*	2007	23	71	112	Rabi sorghum areas of Maharashtra, Karnataka, AP, TN, and Gujarat under rainfed condition	200 cm tall, Purple coleoptiles, leaves with white midrib, semi-compact, cylindrical panicle, tolerant to shoot fly, and lodging
CSV 23	2007	28	155	133	Recommended for cultivation during kharif in, AP, Gujarat, Karnataka, MP, Rajasthan, Maharashtra, & TN	220 cm tall, shown distinct superiority for fodder yield, resistance to grain mold, rust, anthracnose and zonate leaf spot
*DSV 6 (CSV 25)	2009	34	133	108	Kharif in, AP, Gujarat, Karnataka, MP, Rajasthan, Maharashtra, and Tamil Nadu	Tan plant, medium thich juicy stem, compact cylindrical panicle, glumes pink in colour, white lusterous medium bold oval seeds. Resistant to grain mold with tolerance to downey mildew
B. Forage sorghum						
CSH 20 MFMulti-cut Hybrid	2005	742	198		All forage sorghum growing areas of the country	Tan, dark green heavy foliage with green midrib. Medium thick juicy stem, resistant to foliar diseases
CSV 21 F Variety	2008	378	114		All forage sorghum growing areas of the country	Tan, dark green heavy foliage with green midrib. Medium thick juicy stem, resistant to all leaf spot diseases like grey leaf spot, zonate leaf spot, shooty stripe, anthracnose and downy mildew and tolerant to stem borer and shoot fly. This is also tolerant to lodging and shattering
CSH 24MF Multi-cut Hybrid	2008	913	231		All India Northwest Plain Zone (Zone I including Uttarakhand, UP, Gujarat, Haryana, Punjab and Delhi)	Plant height 200–210 cm, tan, light green foliage, green mid-rib, Narrow cylindrical and semi compact panicle, white medium bold seed, medium maturity, grain mold tolerant
C. Sweet sorghum						
CSV 19SS (Variety)	2005	359	8742	120	All sorghum growing areas of the country	Tan, purple coleoptiles, dull green midrib, pearly white medium seed, high green cane yield, juice yield and tolerant to shoot fly

Contd.

Name of cultivar	Year of release	Grain yield (q/ha)	Fodder yield (q/ha)	Duration (days)	Area of adaptation	Salient features
CSH 22 SS (Hybrid)	2005	340	8704	120	All sorghum growing areas of the country	Yellowish green stem has one visible long notch at bud initiation site. Good cane and juice yield. Moderately tolerant to shoot fly
CSV 24 SS (Variety)	2009*	391	14833	120	All sorghum growing areas of the country	Yellowish green stem. Internode covered, Non-tillering with drooping leaves, grayed orange seeds that are elliptical and bold. Good cane and juice yield. Moderately tolerant to shoot fly

*Notification awaited

Research Strategies

Rabi sorghum: Both high grain and fodder yield under receding moisture situation are essential requirement. The resistance to shoot fly, charcoal rot, drought and cold are important for adaptation in rabi. Bold, round and lustrous grain and higher flour recovery add to the consumer acceptability. Research on rabi sorghum to enhance productivity requires gene pools, breeding lines and parental lines with different adaptation niches from those of kharif sorghum, higher dependence on rabi adapted genetic variability, specific emphasis on grain quality and fodder yield, evolution of high yielding cultivars with maturity duration suiting to different growing conditions defined by soil depth, water retention capability and nutrient use efficiency and situation-specific crop production management techniques which may facilitate yield optimisation under varied nutrient use soil moisture regimes. Possible development of genetically engineered cultivars with resistance to shoot fly may offer opportunity to advance rabi planting and avoid problems associated with terminal moisture stress and low temperature. Due to paucity of complete data base on several aspects of rabi sorghum such as its genetic variability, limits to dry matter production and partitioning, yield components and those which are easily manoeuvrable for achieving quick yield improvement, gene pools expressing higher levels of heterosis, breeding behaviour for grain quality, biotic and abiotic stresses, control and correction of low temperature induced seed set and production problems, nutrient use efficiency among genotypes and under variable receding soil moisture regimes, etc. It is important that such data base is generated to strengthen the rabi sorghum research strategy. The immediate option, however, is improvement of rabi sorghum productivity with the available scientific understanding breeding and production management. Due to high demand for rabi grain for food use and better profitability to the producer, this research demands high priority, additional manpower in relevant disciplines and expanded infrastructure commensurate with the task and time frame(Seetharama *et al.*, 2007).

Kharif sorghum: Even with production potential of 4.5 tons grain/ha achieved with only moderate inputs, the highest state average yield is only 1406 kg grain/ha in Maharashtra and still less elsewhere. Therefore, further genetic enhancement for productivity may have little relevance in the immediate future without high degree of resistance to grain mold. Research on augmentation of economic benefit from the achieved productivity potential by preventing grain mold damage and yield loss from major pests, expansion of demand for the grain for food and alternate uses and increasing profitability from kharif sorghum is to precede further increase in productivity. In the absence of the latter strategy, the decline in kharif area would continue. Incorporation of grain mold resistance as well as resistance to major pests in high yield background could be achieved by combination of conventional or biotechnological breeding. Expansion of demand for grain could be achieved by rendering it mold free, by introduction of processing technology and through promoting its use as feed and industrial raw material for starch and alcohol production. Increase in profitability to the farmer would be possible by a combination of demand expansion, increased productivity at farm level at low unit output cost, value-addition to the crop with sweet stalk and high biomass yields and favourable public policies. Specialised kharif sorghum farming for grain export to international market is another emerging option.

Industrial sorghum: Sorghum, particularly kharif sorghum with high productivity potential, has capability to emerge as a commercial crop of industrial importance. With progress in the overall economic well being of the nation, empowerment of households below poverty line with adequate purchasing power, growth in dairy and meat production and shift in the food habit to increased intake of animal-derived food, a phenomenal demand for cheap coarse grain is expected to emerge in the country. Sorghum along with maize would be favoured coarse grains. Suitability of grain to produce good quality potable alcohol may persist as long as it continues to be a competitive raw material. The chances are brighter as the policy of blending ethanol with petrol as biofuel. The molasses, existing as a raw material will not be able to cater the total raw material requirement for biofuel production alone as the expansion programme is envisaged in more than 10 states in the country. We can introduce this crop as a supplement to sugarcane for its utilisation as a alternative raw material for ethanol production.

Fodder and forage sorghum: During the last 30 years the role of sorghum as a major source of fodder has not diminished while its importance as a forage crop has increased. While there is a scope for dual purpose cultivars in certain regions of the country in the immediate future, growing dairy industry may require cost-effective and efficient fodder or forage production system for supply round the year. This offers a vast scope for developing high yielding, good quality cultivars for forage and similar cultivars for fodder supply. As fodder production is slowly becoming commercial, the dual purpose cultivars may be found yield limiting and this may shift the option to fodder sorghum cultivars. Apart from biomass yield, sugar content, protein content and digestible dry matter may assume importance in cultivar development.

Food-Fodder Targeted Research Strategy

Sorghum research in the past was exclusively aimed at increasing productivity and yield augmenting. This was largely justifiable in the context of national need for producing more food grains. Sorghum production situation in this country differs from that in most other parts of the world for the following features. It is being done with sole emphasis on its use as food and hence cultivars with white grain, low tannin content, attractive shape and size and soft to medium endosperm texture are preferred. Little attempt made on nutritional improvement of grain was also aimed at this objective. On the other hand, much of the production elsewhere is used as feed and hence has fewer hassles for grain appearance, texture, etc. The second feature is that sorghum stalk is valued almost equal to the grain as a popular fodder. Hence, sorghum yield implies not only grain yield but also the fodder yield. Biologically, these traits are negatively related.

Poultry industry: This industry, which is growing at an annual growth rate of about 8% in case of layers and 15% in case of broilers, largely depends on maize currently that constitutes 30–35% of poultry ration. The rainy season sorghum has the ability to replace maize in poultry feed to a great extent because of its low price compared to maize (lower by 7 to 37%). At present with the inclusion rate of 12–15% in the poultry feed, about 1.3 m t of sorghum is required by the poultry industry. It is believed that with better qualities of sorghum and improved information dissemination, the inclusion rate of sorghum in poultry feed may be increased to the tune of 25% and hence the increase in sorghum demand.

Dairy industry: With the average consumption of 200–250 grams of grain per day per dairy animal, approximately four million tons of dairy feed is required and the same was produced by commercial sector during 1998–99. Based on the prevailing trend and at an average inclusion rate of 10%, about 0.6 million tons of sorghum is required to produce around 6.04 million tons of dairy feed by 2010 AD. Some giant co-operatives in Gujarat were using sorghum grain at an inclusion rate of 20%.

Alcohol industry-Potable alcohol: Low cost of molded grain is as on taken advantage for its use in the production of potable alcohol. Quality of this alcohol is good and hence may find ready foreign market for blending purpose compared to that produced from molasses which is not used for potable purpose in developed countries. Some of the sorghum cultivars are closer to barley in their ability to produce malt. This property offers potential for breeding high yielding cultivars for malting. Sorghum malt is cheaper than barley malt and would be acceptable for lager beer brewing and production of malt syrup, baby food and non-alcoholic beverages. There are about 295 alcohol distilleries in India with installed capacity of 3198 m litres of alcohol. Nearly 52% of total alcohol production is used for potable purpose (Indian Manufactured Foreign Liquor-IMFL or the country liquor) and remaining 48% is used for industrial purpose. The demand estimates for potable alcohol may drastically increase as MITCON, Pune is active in advocating of potable grain alcohol from sorghum whose feasibility projects are under review. A proposal on diverting kharif sorghum (rain damaged) for beer/alcohol making has been made to Government of India. If approved, this may create a demand for 4 million tons of kharif grain. The decision of Government of India to blend ethanol with petrol will tremendously increase the requirement of anhydrous alcohol in the near future and keeping this in view, the potential of sorghum as a raw material for alcohol production can be greatly exploited. If the government policies on grain based alcohol production become more favorable, there is a possibility of expected growth beyond 10% in case of potable alcohol, as a result of which the demand for raw material (sorghum grains) will be around 0.25 to 0.35 m t by 2010 AD (Seetharama *et al.*, 2007).

Starch industry: The currently preferred raw materials are maize and cassava. The starch content and quality are similar in sorghum and maize has the advantage of maize grain size, high extraction and return from bye-products like corn oil and residue. Specifically bred sorghum cultivars with larger grain size and starch extractability can promote the prospect of sorghum as a competitive raw material in this industry. Demand for industrial starch and starch bye-products is continuously rising. The demand for sorghum in various industrial quarters in a nutshell is depicted in Table 5.2. The revised demand by industries for sorghum shows that the poultry feed industry (especially layer feed) is going to the major buyer which will absorb huge quantity of sorghum, followed by dairy feed industry. It also shows that alcohol industry too will place a fair demand for sorghum. But the starch industry may zero its demand in the future.

Feed industry: Poultry feed industry in the country is progressing well. Several independent feeding trials conducted on layer and broiler birds could conclusively demonstrate that sorghum is as good as maize although, the latter is marginally superior in total energy value. Sorghum based feed in layer birds has to be supplemented with carotene to ensure yellow colour to egg yolk. With the current price advantage over maize, sorghum is used more and more in poultry feed.

Table 5.2. Summary of industrial demand for sorghum in India ('000 tons)

Industry	Estimates[3]	
	2001–2002	Projected for 2010
Poultry feed (Total)	1078⁺–1270⁺⁺	2668⁺–3085⁺⁺
Broilers	240	840
Layers	768 – 960	1668–2085
Others (grower & parent stock)	70	160
Diary feed	440 – 450	600
Alcohol	92	216
Starch	–	–
Exports	11	–
Total	1621–1823	3484–3901

Note: 1. The poultry and starch industries use sorghum only when maize is expensive or not readily available. 2. Figures are projections. 3. Figures are the estimates based on industrial survey conducted as a part of RNPS-6 NATP projects. Figures reflect the average sorghum utilization during 2002 based on inclusion rates and current requirement of raw material. The starch industry figures are not readily available. The annual growth rates considered for the projection are 8% in layer; 15% in broiler; 5% grower and parent stock, 3.5% in dairy, and 10% in alcohol industry. Sorghum inclusion rate in feed considered for the estimation are 12%⁺ to 15%⁺⁺ in layer; 10% in broiler; 5% in grower and parent stock; 10% in dairy.

Malt: Another potential area could be the use of sorghum malt and as an adjunct in brewing industry. Malted sorghum is used for brewing beer in Ghana and Nigeria. Sorghum's comparative advantage would be its low output price, especially in production regions of rainy sorghum in India over that of existing raw material.

Sorghum for bio-energy harvest (bio-fuel production): Sweet stalked and high energy sorghum has the potential to emerge as a major bio-energy crop for production of industrial alcohol, gasohol and even electricity. While the national trials on sweet sorghum had shown its potential productivity as 50–60 t stalk and 2–3 t grain/ha in 125–130 days, reports from China indicate a far high productivity to the tune of 90 t stalk and 6 t grain/ha from the crop in 140–150 days. This productivity potential places the crop at an undoubtable advantage over sugarcane or molasses at their ruling prices for ethanol production. As a bio-energy crop sweet sorghum is more efficient than sugarcane because of its high productivity with relatively low levels of fertilisation and irrigation.

A future policy to blend ethanol to gasoline at 10-20% proportion may create enormous demand for alcohol which could not be met from the indigenous molasses production. Sweet sorghum, in this context, could be effectively exploited as an energy crop. The economic benefit to the nation is saving of more than 10% of the annual fossil fuel import bill. This would grow into a huge saving when the energy use raises many-fold with economic growth from its current low per capita use. A part of the kharif area having most favourable soil and rainfall could be appropriate region for growing sweet or high energy sorghum. The grain of value added high energy sorghum could become a most competitive source for feed or industrial uses.

Production of jaggery and brown or colourless syrup or high fructose syrup are other opportunities from sweet sorghum. These diversifications could be done at village or farm level as small scale enterprises. These agri-businesses possible from sweet sorghum may bring in increased employment and economic growth in the traditionally poor sorghum growing region.

Sorghum markets: Gradual commercialization of Indian agriculture and integration with the world markets on one hand, and decreasing consumption as food, and only slowly increasing demand for sorghum for other uses, an erstwhile subsistence crop, assumes great importance. Although significant amount of sorghum is traded, markets are not well developed for value added and ready to use products are needed to popularize sorghum as a food. While industrial liasioning and understanding the market players behaviour and channels for promotion of alternate uses like sorghum grain for potable alcohol and sweet stalks for fuel alcohol can auger good prospects for sorghum. The increasing emphasis on sorghum fodder calls for appropriate shifts in emphasis to different criteria for advancing materials in the AICISP programme, and the associated regional research and policy imperatives.

Increasing export possibility: The USA, Argentina and Australia are major sorghum exporters, with USA dominating with 72% of trade. Japan and other Asian countries like Korea, Taiwan, countries in the Gulf region, Venezuela and Mexico are major importers of sorghum. The demand for sorghum is expected to be fuelled by rising demand for livestock products following economic growth in the developing countries, change in animal production from bovine meat to monogastric animals such as poultry and swine and with the emergence of intensive grain-based livestock production. While developed countries are expected to maintain their dominance in the coarse grain export market, almost on par with their primacy in the production and trade of wheat, the unfolding global scenario on coarse grains, especially sorghum and maize, allows opportunity for the developing countries to expand their export capability and competitiveness(Seetharama *et al.*, 2007).

The other commercial avenue that Indian sorghum can find is export. The scope of choice of post-rainy season sorghum for export is limited as it is a valued food grain in India itself. Thus the rainy season sorghum can be targeted. The sorghum exports from India have shown fluctuations. Usually sorghum (other than kharif hybrids) are exported to countries like Saudi Arabia, UAE, Belgium, Canada, Kuwait, UK etc., During the year 2002, 8930 metric tonne of sorghum were exported from India.

According to a study conducted by NRCS, Indian sorghum was not found to be competitive in International market. It shows that the Nominal Protection Coefficients (NPC) of Indian sorghum exports under exportable hypothesis were 1.45 for AP, 1.42 for Maharashtra and 1.93 for Karnataka which are prosperous sorghum growing Indian states. On the other hand, if sorghum is imported, the domestic sorghum price will come down by 9% which would potentially affect the farm harvest price in the long run. The research towards the grain quality upgradations, lowering production costs and providing better storage and transportation could definitely find a place for Indian sorghum in International markets which would enhance its export competitiveness in the global sorghum market (Seetharama *et al.*, 2007).

Coverage Strategies *vis-à-vis* Impact Assessment

In the initial years of sorghum R&D, major impact on productivity, production and utilisation could be felt. This impact, however, was dissimilar in kharif and rabi sorghum as well as the crop grown in different agro-climatic regions during the same season. Public policies on sorghum production and utilisation *vis-à-vis* those on "fine" cereals, and on constraints of commercialization of dryland food crop increasingly disfavoured sorghum. Thus, the significant impact and the wider application of the new production technologies of sorghum are showing declining trend(Seetharama *et al.*, 2007).

Kharif sorghum: Varietal improvement was the prime mover of kharif productivity during the last three decades. Advances in productivity achieved from the first commercial hybrid, CSH 1 to the more recently evolved hybrid, CSH 16 and later ones are significant both in respect of grain and fodder yields and in the levels of genetic tolerance to many of the biotic stresses. Kharif sorghum productivity rose from 546 kg/ha in 1970 to 957 kg/ha in 2003, reflecting a 75% increase. Compound growth rate (CGR) of productivity achieved a phenomenal 3.55% during the period 1967 to 1980. This CGR surpassed those of all other dryland crops (and even rice and wheat) during this period. The CGR of kharif sorghum's productivity for the period 1967–2005 and in other sub-periods is lower than that of rice and wheat for the corresponding period. A recent growth trend on kharif sorghum's productivity is very much higher than rice and wheat for the period 2000-2005 (Fig 5.2d). Though kharif sorghum is being pushed to the marginal environments, the recent productivity performances were much better than the past. This may be due to replacement of earlier hybrids with CSH 16, CSH 18 and CSV15. The HYVs, specifically the hybrids, which were the major causal agent for the productivity growth and carrier of production technology found steady and good acceptability among farmers so that with least developmental support they could cover nearly 94% of the cropped area.

State-wise HYV coverage *vs* productivity performance (combined kharif and rabi): The coverage of HYV in states such as Maharashtra is almost 100%. The predominant influence of HYVs in pushing up productivity is evident from the relationship between the coverage of HYVs and the productivity levels of different states. On account of HYVs spread, particularly the hybrids, there was an emergence of a dynamic seed production and trade network for sorghum. This trade interest also played a significant role in the spread of these hybrids. One important factor which has to be reckoned is slower pace of replacement of the earlier popular hybrids with that of more recent ones. It may be due to poor extension efforts, or poor performance on the farmers' fields, and low crop management. Thus, only few HYVs officially released could see the light of the day and found acceptance by the farmers.

In general, increase in yields of HYVs contributed to increase in total production, leading to its stability and security if the crop in question is a staple cereal crop. Analysis on HYV coverage reveals that in 1972, the gap between total area under sorghum and area under HYVs of sorghum was wide which has narrowed in 1998 (to be updated), thus making both the total area under sorghum and area under HYVs almost equal. It appears that CSH 9, released 20 years ago, still is the kharif dominant hybrid among the public bred ones. The research and extension efforts for introduction of HYVs were carried out well over the years. But the real problem is that the recently released HYVs are not able to penetrate or replace the earlier ones though superior, which may explain to some extent why the yield growth rates are staggering of late. Indirect impact of HYVs has lead to sparing sorghum area to other commercial crops such as pulses and oilseeds as the same amount of sorghum was produced as before from reduced land under declining or constant demand as a staple food. Such replacement has lead to increase in standard of living of dryland farmers due to significant rise in their income levels.

Growth in productivity was also aided by the production technology packages perfected for the popular HYVs. Notwithstanding that these packages were not wholly adopted and the extent of their adoption was highly variable under different agro-climatic regions, the very cultivation of HYVs stimulated application of major fertilizer nutrients. Better response of HYVs to these nutrients also helped in pushing up the yield level with nutrient management. Response to N in HYVs, on an average, is 20.8 kg grain/kg N at N application up to 40 kg/ha in comparison with about 6 kg grain/kg N in local cultivars. The HYVs have shown a cost-benefit ratio of 0.73:1, offering a 27% gain on investment in fertilizers. Compared with other dryland cereals and maize, sorghum has higher nutrient use efficiency. The HYVs, in general, are not superior to the traditionally grown land races in biomass productivity, but are far more efficient over the latter in partitioning dry matter into grain. Hence, the overall nutrient depletion from the soil with continuous farming of HYVs is no way significantly different from the earlier farming situation with traditional varieties. This helps to maintain the production sustainability of sorghum farming soils.

Compared with the landraces, which had better adaptability under specific agro-ecological conditions, the HYVs have lesser resistance to pests and diseases. While the shoot fly resistance in kharif HYVs is not to the desired level, significant yield loss from this pest damage is obviated with the recommended early planting soon after the onset of monsoon. Late planted kharif HYVs are seriously prone to this pest. Stem borer is similarly a serious pest to HYVs, particularly in the north India. The level of resistance to other pests such as head bug, aphid and shoot bug is no better. Economic loss from these pests, fortunately, is not very serious when planted early. Notwithstanding a significant success achieved in identifying multiple sources of resistance against these pests and incorporation of some of these sources in the present day HYVs, much remains to be done in this respect. In general, the high yielding varieties in comparison with the hybrids are found superior in genetic resistance to major pests and diseases.

The HYVs, however, with their reduced maturity (selected to escape terminal drought) became more vulnerable than the traditional cultivars (filling grains after cessation of rains) to grain mold damage as the grain development stage of HYVs coincides with rainy and humid days. This leads to varied extent of mold damage. Depending on the grain developmental stage when molding occurs and seriousness of the incidence, the loss could be complete or partial. Consequently, the grain may become pink in colour with complete damage of discoloured to black under prolonged wet conditions. A molded grain command much less demand, cannot be stored long, and fetches lower prices. The economic loss from grain molding is most serious and is the primary cause adversely affecting the competitiveness of kharif sorghum. No data are available on the economic loss to the nation on grain molding. Chemical control of molding is either too costly or ineffective. Genetically resistance, as understood, is hard to achieve along with desirable white grain and texture. Regular incidence of molding of kharif grain in association with differences in the public policies on the production, procurement and distribution of fine cereals *vis-a-vis* sorghum, seriously limited an assured return from a good crop, and in turn the demand for kharif sorghum grain. Some loses can be contained by harvesting grain at physiological maturity and drying (Seetharama *et al.*, 2007).

Seed Systems

Formal seed supply systems: Seed supplied through the formal seed sector is characterized by planned production, some form of (mechanized) processing, inclusion of only identified/notified varieties and a system of quality control. These seed units generally deal with large quantities of seed. They correspond to "organized seed sector" (Cormwell *et al.*, 1992, Tonapi and Seetharama, 2004a, b).

The formal seed supply systems consist of:

(i) Seed production by the national government agencies,

(ii) Seed production by the state government agencies,

(iii) Government-assisted and other co-operatives,

(iv) Multinational (MNCs) or transnational companies (TNCs),

(v) The domestic private sector:

  (a) with its own R&D

  (b) without its own R&D.

(vi) Joint ventures such as:

  (a) MNC with a domestic private company,

  (b) A domestic company with another domestic company, etc.

The formal seed sector—through its involvement in variety development, evaluation and release; seed production, marketing and distribution; and quality control operations—can offer only seed of a limited range of varieties and within specified quality standards. In practice these may constrain the ability of the formal system to meet the diverse needs of farmers in project area. There are serious concerns on the appropriateness of the varieties available, quantity and quality of seed delivered, seed production costs and prices and timeliness of supply. More important, rigid government policies, regulations, poor organizational linkages and inadequate infrastructure contribute to the problems of the formal system in developing countries. Moreover, the existence of even a relatively developed formal sector at the national level certainly cannot guarantee small-farmer seed security at community and household levels. Private seed companies are reluctant to produce and market varieties of sorghum, pearl millet, pigeonpea and groundnut because the varieties do not fetch good prices. Even if they did, they might not reach small farmers in remote areas. Moreover once varieties are sold to a farmer, he would save his own seed for the next season, and not buy again, thus nullifying any interest the private seed company may have.

Informal seed supply systems: Informal seed supply systems are characterized by lack of functional specialization; they are heterogeneous in space and flexible in time. These systems are traditional, informal, operating mainly at the community level through exchange mechanisms and involving limited quantities per transaction (Cromwell *et al.*, 1992).

These systems broadly include:

(i) Farm-saved seed and farmer to farmer exchange,

(ii) Farmers' co-operatives,

(iii) Community groups,

(iv) Seed growers' associations,

(v) Non-governmental organisations (NGOs), etc.

Interventions Needed in Local Seed Systems

Various formal and informal approaches are applicable to seed supply. In fact, what is encountered is usually a complex mix of the two in varying proportions. Hence search for the alternatives to promote the availability of best quality seed to the farmer at the right time and at the right price becomes the prime issue that needs viable intervention. In such a perspective, the components of the seed supply system would emerge as neither necessarily complementary nor supplementary in relation of one to the other although it may sometimes be so. Various components in the chosen mix would assume equal importance in approach but not in their relative contribution toward seed supply. Further, no components may probably exist as mutually exclusive; several alternatives may exist and function together. Therefore, the following interventions and future strategies could be considered to strengthen alternative seed systems (Ravinder Reddy *et al.*, 2006):

- Strengthening different *sui generis* amalgams and their components that may operate as per the need of a crop, situation or region.
- Farmers' participatory seed production, on-farm trials, demonstrations, on-station selection by farmers and auctioning for private supply systems.
- Building self-reliance of farmers for seed availability through such activities as '*Beej Swavlamban Yojana*' (seed self-reliance programs).
- Decentralized cluster approach through the seed village program.
- Meeting infrastructural and contingent needs through the seed bank scheme.
- Strengthening of seed-growing contracts.
- Making farmers' seed cooperatives the usual parastatals.
- Cautious extension of facilities, providing a level playing field, wherever appropriate. Expanding the seed village concept and strengthening of

seeds in forage crops, medicinal plants, flowers and underutilized crops.

- Setting up community seed banks for seed exchange, distribution and utilization.
- Setting up *in situ* conservation centres for preservation of varieties.
- Evaluation, characterization and multiplication of these varieties involving the community, namely farmers.
- Survey, collection and documentation of indigenous grain varieties.
- Encouraging farmers to grow these indigenous varieties.
- Setting up farmer self-help groups to maintain and run community seed banks and drive the seed conservation effort.
- Developing and strengthening self-devised marketing support and self-generating credit support.
- Creation of controlled conditions for effective nuclei in seed production and development of processing, storage and packaging facilities.
- Encouraging public-private sector partnership and development of methods for mutual learning between local people and formal scientists.
- Traditional systems and practices can contribute to the preparation of village micro plans in relation to sustainable seed systems. The plans should be based on both geographic and traditional community boundaries rather than only on administrative boundaries.
- Revival of the traditional seed management systems that served the society for hundreds of years but are currently threatened.
- There is a clear need to integrate traditional and formal systems for participatory monitoring and taking feedback to achieve adaptive strategies for management of community seed systems.
- Variety diffusion into local seed systems is more crucial than its formal course. The products in the form of varieties or parental lines of hybrids, etc. could be distributed through minikits.

PRODUCTION AND MARKETING OF SORGHUM HYBRIDS

Salient features of sorghum seed production and marketing

(i) Seed production of the sorghum hybrids is taken up during the rabi season only (mainly because of mold threat during kharif). The Andhra Pradesh state meets 80% of the total sorghum seed requirement. Part of the seed is produced in Karnataka state also.

(ii) The area under sorghum seed production during the period 1997–

2004, shows significant decline in the seed production area under public bred hybrids. Private sector has produced a major part of the seed required by the industry during the period of fast decline of public sector seed industry.

(iii) While the area under private bred hybrids increased two fold till 2003, the total area under private and public bred hybrid seed production has almost halved in each case in 2004, indicating a decline in the overall demand of the sorghum hybrid seeds. Similar trends are seen in the total quantity of seed produced during the period 1997–2004.

(iv) The reduction in the area under commercial cultivation of the sorghum crop especially during the rainy season has a direct bearing on overall reduction of the seed quantities produced. Hence there is a pertinent need to increase the seed production and uptake of public bred sorghum hybrids.

(v) Creating demand for grain and substantially bridging the gap between cultivar production potential and farm yield with the aid of cost-effective production technology and marketing support can only arrest the trend in decline of sorghum seed production. Industrial utilization of sweet stalk can bridge these gaps to increase seed production and profitability, provided sweet sorghum technology can be popularized. Demand for forage seed is enormous but margin of profit is very small (0.50 paisa/ kg to the farmers).

(vi) Market size of the Indian seed industry is estimated to be of worth Rs 3500–4000 crores. Vegetables seeds have a major share with 19%, followed by 13% each of cotton, sunflower and wheat. Sorghum accounts for 7% of the total industry. Out of the total area of 9.2 million ha under sorghum cultivation, hybrid seeds occupy 22% with a value of Rs 70 crores. The most potential sorghum seed markets are Maharashtra (7800 tons), Karnataka (2400 tons), Madhya Pradesh (800 tons), Andhra Pradesh (720 tons), Gujarat (240 tons), Rajasthan (200 tons), and Tamil Nadu (400 tons).

(vii) The emerging potential seed markets are Maharashtra for summer sorghum (100–150 tons), and Karnataka for summer (150–200 tons), and Andhra Pradesh (200 tons) during post-rainy season (and recently even in rice fallows wherever water supply is declining) which is raising some hope for seed industry.

(viii) The major players in the sorghum industry are Mahyco, Emergent Genetics, Proagro, Pioneer, Advanta, Bioseed, JK Agri Genetics Ltd., Gangakaveri, Zuari seeds, Nuziveedu Seeds, Ajeet Seeds, Nath Seeds, etc.

Size and Scope of the Sorghum Seed Industry

Both public and private sector organizations meet the requirements of the Indian seed industry. Public sector contributes to the tune of 41% and the large private sector companies contribute 33%. Small companies and unorganized segment contribute 13% each.

The market size of the Indian seed industry is estimated to be worth Rs 3500 to 4000 crores. Vegetables seeds have a major share with 19%, followed by 13% each of cotton, sunflower and wheat. Sorghum accounts for 7% of the total industry. The crop wise area and market size is given below (as in year 2003–04). Out of the total area of 9.2 million ha under sorghum cultivation, hybrid seeds occupy 22% with a value of Rs 70 crores (Ramakrishna *et al.*, 2006).

Crop Wise Area and Seed Market Size in India

Seed production of the sorghum hybrids is taken up during the rabi season and primarily in Andhra Pradesh, meeting 80% of the total industry's requirement. Part of the seed is produced in Karnataka state also. When we look at the area under sorghum seed production during the period 1997–2004, we can see the steady and significant decline in the seed production area under public bred hybrids. While the area under private bred hybrids increased two fold till 2003, the total area of both private and public bred hybrid seed production almost halved in each case in 2004, indicating a decline in the overall demand of the sorghum hybrid seeds (Ramakrishna *et al.*, 2006).

Crop wise area and seed market size in India

Crop	Area (million ha)	Hybrid (%)	Volume (quintals)	Value (Rs in crores)
Sorghum	9.2	22	140,000	70
Pearl millet	7.6	37	140,000	84
Maize	6.0	41	310,000	71
Rice	44.0	1	65,000	60
Wheat	24.0	—	10,000	6

Contd.

Crop	Area (million ha)	Hybrid (%)	Volume (quintals)	Value (Rs in crores)
Cotton	7.8	52	100,000	350
Sunflower	2.4	85	100,000	180
Castor	1.0	60	30,000	30
Total field crops	102.0	—	895,000	851
Tomato	0.8	34	340	40
Bhendi	0.6	21	6,500	40
Chillies	1.0	4	100	28
Others	0.6	—	140	100
Total vegetables	2.4	—	6,940	208
TOTAL	**104.4**	**—**	**901,940**	**1059**

Similar trends are seen in the total quantity of seed produced during the period 1997–2004. Private sector has produced major share of the seed required by the industry whereas the public sector share has steadily declined. The reduction in the area under commercial cultivation of the sorghum crop especially during the rainy season, has a direct bearing on overall reduction of the seed quantities produced.

The main reasons for significant reduction in the kharif area are:

1. Increased productivity enabling release of area for other crops
2. Changing food habits
3. Poor quality of the grain resulting in low price
4. Increased profitability through commercial crops
5. Most of the area taken by cotton and soybean.

Sector Wise Market Size of Seed in India (Relative Percentages)

Sorghum Hybrid Seed Production and Profitability

Hybrid seed production of sorghum is a specialized operation that needs higher level of competence to meet quality requirements unlike commercial grain crop. The cost and the net returns of sorghum hybrid seed production provide a scientific basis for fixing compensation to the seed growers. The estimated variable cost and fixed cost of sorghum hybrid seed production, respectively, amounts to 81% and 19% of the total cost of seed production. About 57% of variable costs are due to the human labor and material inputs. The net benefit: cost ratio works out to 0.45 under normal seed production scenario. The total net profit is to the tune of Rs 11,775/- if seed production is taken up on own land, and on leased land, the seed profitability is to the tune of Rs 4525/- per acre (Ramakrishna *et al.*, 2006).

Hybrid seed production–cost & returns (ac^{-1})

Item	Cost	Item	Income (Rs)
Seed	500–1000	Seed yield	12000–18000
Land preparation	1000	(8–10 q/acre)	(@1500–1800/q)
Inputs	MALE YIELD	2250–4800	
Fertilizer	1000	(5–8 q/acre)	(@450–600/q)
Pesticides	500	FODDER	500
Irrigation	1000		
Labour	3000		
Land lease	6500–8000		
Total cost	Total income	14750–23300	
Net profit	Own land	Rs 11775/acre	
	Leased land	Rs 4525/acre	

Economic Viability

The size and scope of the sorghum seed industry centered around Hyderabad is very large, and there is fierce competition. However, with the kharif grain area going down with inconsistent and low market demand (low price offered; support price is on paper–there is no procurement), the morale and business is down(Vilas Tonapi, 2009).

Further,

1. Most of the seeds business is for kharif hybrids only. Forage hybrids are few and only few companies work. There is no significant forage seed production during kharif.
2. Most seeds sold are only truthfully labeled (very little certified seed).
3. Some major multinational companies have closed sorghum options since last year, and the marketing scene is dominated by a large number of small and niche players.
4. Farmers are real seed producers in most cases, and traders buy seeds from them and market as truthfully labeled seed.
5. Public corporations are least effective. States like Karnataka advertise and allow private companies to sell with state subsidy.
6. Mahabeej Maharashtra State Corporation, till recently the major producer of sorghum seeds, is like a company now, and will sell only its own branded seeds.
7. About 80% seed production is concentrated in Nandyal area. Nizamabad dominates in forage seed production. Forage seeds are specially sold in truck-units (packed in ordinary gunny bags only) and mostly through informal channels.
8. Significant differences in cost of seed production across locations, and especially hybrids are noteworthy. CSH 14 is cheapest to produce because: (i) good synchrony between parents for time of flowering, (ii) quantity of hybrid seeds produced is high, and (iii) male parent also gives high yield of premium quality grain.
9. *Lesson from the above:* Our release procedures need to be stringent. At this time, we are mostly taking the word of breeder (what appears in proposal) at face value, and are not tested; *testing in seed production zone under realistic conditions, and not the data collected at station, should be relied upon.*
10. "Seed village" concept operates for forge hybrids in Nizamabad, but not under any contract (or regulation).
11. Availability of water during off-season is the major limiting factor for hybrid seed production. Insects are controlled effectively, and so also

diseases. (Ergot is not a threat, but needs constant monitoring).

12. Viability of sorghum seeds is low, so seeds have to be sold in the same year it is produced. This aspect deserves much attention of researchers.
13. Minimum quantity of certified seeds to be produced and sold for an industry to be viable: 200 tons of seeds per year.
14. State governments such as AP have stringent control on seed quality; in spite of spurious seeds enter the market.

Popularization of Public Hybrids

Technology transfer is generally slow and tardy in the country due to the multitude of farm holdings involved in the production system and multiplicity of agencies for the transfer. Rapid percolation of technology to the targeted production areas is an essential requirement. When the production advance is primarily driven by HYVs, the seed becomes mother of all technologies. Production and distribution of quality seeds had often stood in the way of rapid coverage of cropped area with high productivity technology. The private sector industry had played signal role in this respect in the past. The emerging scenario with plant breeders' rights on new cultivars and private sector having its own parallel varietal improvement research, the responsibility of transfer of patented technologies would be left with the patent holding institution or its authorized agent. This may sieve technologies for their efficacy and competitiveness and those superior alone would gain acceptability. In public funded research, public policies on technology transfer may also have bearing. New information technology regime with websites and internet system is also expected to play some role in technology transfer.

The current PRA and RRA techniques in rabi sorghum are of immediate relevance to understand farmer's mood, farmer traditional technology with variable seed rate and plant density and farmers reasoning to do so. Farmer—NGO Scientist participatory approaches, if executed, may help to quickly identify production technologies including genotypes under farmer management for local adaptation and acceptability traits including *roti* making quality, storability and fodder profitability for expeditious spread of HYV and improve zone wise productivity.

Public *vs* Private Partnerships

Problems in seed supply are commonly identified as key constraints affecting the rate and level of adoption of new varieties. They reduce the payoff to national crop improvement programs. For the most part, these constraints relate to market conditions and associated seed production and trade policies. Hence, there is a need to identify options for strengthening seed multiplication and trade based on issues of national seed policy, commercial profitability, and the opportunity to exploit alternative seed supply systems such as NGOs and farmer groups. One component of this work seeks to identify opportunities for strengthening informal seed supply systems including direct seed trade between villages and between farmers. In addition, significant gains in productivity can be derived from linking policies favoring the liberalization of fertilizer markets and strengthening of a more competitive fertilizer trading systems with stronger sets of fertility management recommendations. Several of the identified research objectives will lead to commercializable products. An effective linkage will be established for effective partnership. This will cover effective exchange of pre-breeding material, parental lines, genetic transformation and marker protocols etc.

Actions Suggested

1. Concentrate on IPR–develop and license seed parents, value-added germplasm, and process using new and novel technology.
2. Seed business only as public good: facilitate quality and price control.
3. The sorghum seed programme cannot be economically viable on its own, but can be doubly viable with linkages and partnership with public and private sector. The cooperative and partnership mode will increase its sustainability, economic viability and profitability both to NRCs and partners.
4. Competence: Knowledge and suitable materials exists, but week in marketing.
5. Facilities are inadequate, but can be built up however, it is better to outsource many jobs to local service providers. Both public and private sector organizations meet the requirements of the Indian seed industry. At present, public sector contributes to the tune of 41% and the large private sector companies contribute 33%. Small companies and unorganized segment contribute 13% each.

Suggested Steps for Commercialization of Hybrids

1. In-house testing
2. Company multi-location testing
3. Government trials at state and centre levels
4. On farm trials-strip tests, mini-kits, farmers' selection
5. Test-marketing

6. Product launch and sustaining market share
7. Licensing of the hybrids.

CONCLUSION

Apart from quality deterioration caused by mold, shrinking demand for kharif grain is the consequence of governmental policies. Protection of this grain from mold damage is expected to enlarge its food demand on the singular reason of its low price. While shift from coarse grain to fine grain is a welcome change, the tenuous production-demand equation of food grains in the country does not favour to underwrite the importance of sorghum as food grain. As long as production of fine cereals does not cross the surplus barrier, importance of sorghum, though at reduced level, may persist. Economic prosperity, on the other hand, may fuel the demand for livestock products, which in turn may generate demand for more coarse grains for livestock production. Production for feed in domestic parlour and for international trade and for industrial uses may generate adequate demand for kharif sorghum. This may require cultivar development with specific grain attributes, high productivity to enhance cost competitiveness and other required standards. Mission oriented R&D activities on these aspects involving client institutions, state of the art know how and appropriate governmental policies promoting this production may provide solution to the problems of kharif sorghum.

The low return from kharif sorghum to the producer is due to the low prices arising from low demand for molded grain and an equally low value for the kharif fodder. This is the primary reason for diversion of area from kharif sorghum. There are two possible ways to improve the compatibility of kharif sorghum. First, by creating demand for grain and substantially bridging the gap between cultivar production potential and farm yield with the aid of cost-effective production technology. Second, by value-addition to the crop with sweet stalk. The latter approach may be extendable only to favourable production situation depending on the industrial utilization of sweet stalk. Bridging these gaps will certainly help to increase seed production and profitability. In fact, the most potential markets for seed are Maharashtra (7800 tons), Karnataka (2400 tons), Madhya Pradesh (800 tons), Andhra Pradesh (720 tons), Gujarat (240 tons), Rajasthan 200 tons, and Tamil Nadu (400 tons). The emerging new seed markets are Mahrashtra for summer sorghum (100–150 tons), Karnataka for summer sorghum (150–200 tons), and Andhra Pradesh (200 tons) for post rainy season. The major players in the sorghum industry are Mahyco, Emergent Genetics, Proagro, Pioneer, Advanta, Bioseed, JK Private Agri Genetics Ltd., Gangakaveri, Zuari seeds, Nuziveedu Seeds, Ajeet, Nath etc.

Strategies and Approaches to Commercialize Public-Bred Hybrids and Varieties

To have a successful seed production programme and to regain the past glory of sorghum, we need to locate alternate areas for seed production, develop hybrids with high seed yield potential, retention of area under kharif sorghum, find a solution to the grain molds through biotech tools and strengthen the rabi hybrid development programme. The main underlying strategies to commercialize public-bred hybrids and varieties should involve the following:

1. **Enabling policy environment**
    - Ensuring access to a wide range of varietal technologies, and services which provide seed security
    - Put in place strong public sector breeding program
        - Compete effectively to realize return on investment
        - Effectively manage an entire process of negotiation and licensing
        - Breed innovative hybrids for a wide range of environments and customers.
        - Utilize novel genetic, intellectual and technological resources
    - Developing cost effective and reliable strategies to ensure stocks to meet market demands
    - Timely and cost effective delivery of high quality seeds to the point of their sale
2. **Enabling legal environment**
    - Decentralized licensing service and effective enforcement of standards
    - Put in place the mechanism to protect plant varieties and plant breeders rights under Protection of Plant Varieties and Farmers' Right (PPV&FR) act of 2001
    - Protection of breeders and farmers rights to encourage new varieties
    - First step before privatizing seed production of public bred hybrid is to register the parental material
    - The genetic identity of lines should be supported by molecular data to enable better protection of varieties

3. **Enabling organizational climate**

- Public agricultural research must address the challenges of privatization
- Attention must shift to specific opportunities for interaction with the private sector
- Public plant breeding must become more open
- Seed regulations must encourage private plant breeding
- Establishment of IP management cell
- Freedom for operation
- Enabled IP environment (PVPFR and patents)

Change in mindset

- From regulator to facilitator
- Engage in partnerships
- Skill development and complementation

Legal Framework

- Formation of single window for dispensation in spirit and practice
- New legislations be made facilitative rather than regulatory
- Effective IP protection and enforcement across the board

4. **Enabling the seed industry: areas of partnership**

- Involve private sector in maintenance and production of parental lines, seed production of hybrids, also help them in identification of alternate locations
- Make provision for sharing of research facilities- laboratory, land and personnel
- Provision of research grants and fellowships to public sector by private sector and enabling contractual research
- Training of private sector in seed technology and Bio and other technologies
- Involve private sector in multi-location trials and also in large scale seed production and popularization of cultivars and large scale front line demonstrations and strip plot testing
- Subsidy for private sector hybrids under national food security mission needs to be provided
- Public and private sector are not involved fully as yet. Private sector is very strong in transfer of technology and public sector

is weak. This advantage has to be harnessed through public-private partnerships

- Create coordination/net work cell to enable forming of a consortium with seed agencies for research activities

5. **Formation of consortium**
   - Each partner must bring value to the table with common objective
   - Value creation for both and each partner to have complementary roles
   - This partnership allows private sector seed companies access to the vast germplasm pool
   - The benefits of such a partnership will be evident in the near term in the form of new hybrids in various crops
   - Development of specific parental lines and sharing based on contractual reserach
   - Trait specific product development under joint research
   - Joint research for specific hybrids for a company who invest in the program
   - Joint IP right for the finish product should be put in place
   - Induction of desired trait in the parental line of the company on contractual research mode
   - Benefit sharing mechanisms be given a legal framework
6. **Essential organizational outputs**
   - Standardize seed production technology before licensing
   - Establishing effective distribution networks
   - Economic prices of improved seed
   - Increasing effective demand for improved varieties among smallholders
   - Decreasing the cost of seed production and distribution
   - Improving infrastructure, rules and data on predictable effective demand
   - Standardize regulations
   - Reduce the learning and transactions costs for new seed enterprises
   - Increased role of farmers and NGOs in seed enterprise

7. **Strategies and actions for decentralized licensing**
    - Licensing partners identified and sensitized
    - Identify trustworthy partners and train on licensing
        - Design royalties, terms and conditions
        - Use proven cultivar as bargaining chip
        - Use negotiating skills for a win-win situation
    - Monitor activities of licensing agents
        - Stipulate records that need to be produced annually
        - Evaluate the records periodically to ensure traceability and growth of licensed cultivars
8. **Strategies and actions for a strong public breeding program**
    - Funding increased and operations revamped
        - Prioritize to rationalize allocation
        - Improve conditions of service
    - Resources for research in all priority areas committed
        - Increase funds to research on areas with commercial potential
        - Provide for continuity of research programs
    - Research and market linkages developed
        - Explore industrial uses for all crops
9. **Issues related to PGR and IP: maintaining the flow of germplasm**

    A major policy issue for public sector research organizations is to;

    - Ensure that IPRs do not prejudice the free access to, and flow of germplasm between countries and between programs within a country, and
    - Ensure that growing privatization does not reduce the diversity of varieties deployed in farmers' fields.
10. **Task ahead**
    - Development of an effective model and working out the modalities to optimally harness the strengths of various sectors for a common goal of expeditiously developing and promoting public bred hybrids and commercializing them.

REFERENCES

Cormwell E., Friis-Hansen E. and Turner M. 1992. The seed sector in developing countries: A framework for performance analysis. ODI, London.

Ganga Prasad Rao, N. and House, L. R. 1972. Sorghum in Seventies. Oxford and IBH publishing company 638 pp.

Global Plan of Action for the Conservation and Sustainable Utilization of Plant Genetic Resources for Food and Agriculture, 1996. Leipzig Declaration adopted by the International Technical Conference on Plant Genetic Resources Leipzig, Germany 17–23 June 1996.

Ramakrishna C, Vilas A. Tonapi Vilas A. and Dayakar Rao, B.2006. Sorghum seed production and marketing scenario in India. *In*: Strategies for millets development and utilization. 2006. (*Editors*: Drs. N Seetharama & Vilas A Tonapi). 232 pp. ISBN 81-89335-12-X.

Ravinder Reddy, Ch., Vilas A. Tonapi, Vilas A., Bezkorowajny, PG. and N. Seetharama, N. 2006., Innovations in Seed Systems in Semi-Arid Tropics of Andhra Pradesh, India. 177pp. Fodder Innovation Project, International Livestock Research Institute (ILRI), International Crops Research Institute for semi-Arid Tropics, Patancheru, India and Director, National Research Centre for Sorghum (NRCS), Rajendranagar, Hyderabad (AP) 500 030, India. ISBN 81-89335-07-9.

Tonapi V. A. and Seetharama, N. 2004a. Improved seed systems for small and marginal farms. Paper presented at XVII Annual Conference on Agricultural Marketing, Organised by the Indian Society of Agricultural Marketing (ISAM), and Acharya NG Ranga Agricultural University, Hyderabad Andhra Pradesh, 5–7 February 2004. Abstract in *Indian Journal of Agricultural Marketing* (Conference special), 17: 260.

Tonapi V. A. and Seetharama, N. 2004b. Seed systems for sustained millet productivity. Paper presented in the 3rd National seminar on millets: Research and Development future policy options in India. Jodhpur, 11–12, March, 2003.

Tonapi Vilas A. 2009. Successful producer organizations for a sustainable improvement of seed delivery systems in India. Consultancy report submitted to FAO, Rome. January 2009. 355p.

Seetharama, N., Tonapi, V. A., and Dayakar Rao, B. 2007. Vision 2025. NRCS perspective plan. Indian Council of Agricultural Research, New Delhi, India.

6 Strategies for Promotion of Public - Bred Pearl Millet Hybrids in India

I.S. Khairwal and Sangeeta Kumari*

ABSTRACT

Pearl millet, one of the most extensively cultivated cereals in the world is a principal food cereal cultivated in the drought prone semi-arid regions of Africa and Indian subcontinent and also used as forage, animal feed, building material, fuel for cooking. Commercialization as alternative food, feed, and industrial products is one of the ways to increase demand for pearl millet. At present hybrids cover nearly 50% of the total pearl millet area in India due to high grain yield advantage and rate of out crossing. Requirements, approaches, important considerations for development of parents and hybrids; progress and achievements, alternate areas for seed production, maintenance of parental lines, seed production of hybrids and their management, seed processing, marketing strategies, linkages between public private sectors and protection of hybrids have been discussed.

Key words: Breeding strategies, cultivar options, hybrid, linkages, marketing, seed production

National and International Crop Scenario

Pearl millet is one of the most extensively cultivated cereals in the world, after rice, wheat, maize and sorghum. It is a principal food cereal cultivated in the drought prone semi-arid regions of Africa and Indian subcontinent. India has the largest area (9–10 m ha) under this crop, ranking it third after rice and wheat. In the USA, Australia, Southern Africa, and South America, pearl millet is grown most extensively as a forage crop mainly for animal feed.

In addition to grain and forage uses, pearl millet crop residues and green plants also provide sources of animal feed, building material, and fuel for cooking, particularly in dry land areas. Pearl millet crop has a wide

* AICPMIP, Mandor, Jodhpur - 342 304.

adaptability to local environments. It is hardy and can be grown in areas which are very hot and dry, and on soils too poor for rice, maize and sorghum. Pearl millet is considered more efficient in utilization of soil moisture and has a higher level of heat tolerance than even sorghum and maize.

It is adapted to warm and dry climate but most of the area under the crop is in drought-prone regions of the world. In these areas, pearl millet is usually grown with limited inputs, sparse rainfall and low soil fertility with correspondingly poor yields. The low average yields are primarily a result of the hot, dry conditions where pearl millet is mostly grown, rather than pearl millet plant's capability. Pearl millet hybrids grown under better conditions yield 3000 to 4000 kg/ha compared to 300 to 800 kg/ha when moisture and soil fertility become limiting factors.

A majority of the subsistence farmers in India who typically cultivate this crop are unable to take advantage of high yield potential because they have limited options for improving their management practices. Pearl millet production can be increased through growing hybrids with improved tolerance to drought, diseases and responding to higher rates of fertilizer applications.

It is observed that of late, food habits of pearl millet eating people are changing towards rice and wheat products. Consumer preference for pearl millet has fallen due to irregular supplies, rising incomes and rapid urbanization, particularly in smaller towns. The increasing appreciation of the cost of women's time has encouraged the shift from pearl millet to readily available processed foods (milled rice, wheat flour, etc.) that are far quicker and more convenient to prepare. However, there is hope to improve the farmers' income from pearl millet cultivation in dry area through the adoption of improved cultivars and technologies.

In addition to pearl millet as a food crop, there are possibilities of other alternative uses of pearl millet such as novel foods, processed foods, and as feed for dairy animals and poultry. Thus, commercialization of alternative food, feed, and industrial products is one of the ways to increase demand for pearl millet.

Scope and Importance of Hybrids in Pearl Millet

The protogynous flowering in pearl millet makes it a highly cross-pollinated crop with outcrossing rate of over 85%. The outbreeding system makes open-pollinated varieties (OPVs) as the natural cultivar state of this species. OPVs are cultivated in Africa, parts of India and other parts of the world. *Pearl millet displays high degree of heterosis for grain yield, hence providing hybrids as another cultivar option.* Hybrids are cultivated in India and USA. Although development of OPVs continues in Africa and India especially for drier regions and realizing the benefits of hybrids, development of high-yielding disease resistant hybrids has taken drivers seat in India and USA.

Cultivar Options in Pearl Millet

Open-pollinated varieties: Open-pollinated varieties are developed by random mating of 10–20 selected elite lines (with morphological uniformity) and maintained by random mating, which permits no control over recombination within the populations. The OPV option, however, allows for only partial exploitation of heterosis. Sustainable improvement in grain yield through OPVs faces a major dilemma. On one hand, crosses between morphologically similar genotypes/varieties generate little genetic variability to sustain significant genetic gains, though they permit rapid development of OPVs with acceptable morphological uniformity. On the other hand, crosses between dissimilar types create more heterotic F_1s and large variability in the subsequent generations for sustainable genetic improvement, but populations derived from such crosses also reduce the pace of developing OPVs with acceptable morphological uniformity.

Hybrids: In case of pearl millet, it is the F_1 generated by crossing a male-sterile line and a restorer line for cultivation at the farmers' field. The hybrid route provides an excellent opportunity to use dissimilar parental lines to harness heterosis to its maximum possible extent, and yet produce a cultivar with high degree of morphological uniformity. The breeding system and extent of heterosis, and availability of stable cytoplasmic-nuclear male sterility (CMS) and its fertility restorers, have made the hybrid option commercially viable in pearl millet. A comparison of various aspects of OPVs and hybrids is presented in the Table 6.1.

Table 6.1. Comparison of open pollinated varieties and hybrids in pearl millet

Character	Hybrids	OPVs
Genetic composition	Homogenous and heterozygous	Heterogeneous and heterozygous
Vulnerability to biotic stresses	Vulnerable	Relatively less vulnerable
Adaptability	Narrow	Wide
Maintenance	Crossing two parents	Random mating
Seed production	3 lines required for multiplication	Nothing extra than the OPV itself
	Technically difficult	Technically easier
	Relatively cost intensive	Less cost intensive
Seed cost	High	Low
Seed replacement	Every cropping season	Farmer saved seeds can be used
Seed industry	Require organized seed industry	Require seed distribution mechanisms
Brand name	Provide clear-cut brand names	Does not provide clear-cut brand names
Cultivated in	Well-endowed environments	Marginal environments
Agronomic practices	Required to harvest more yield	Less cared by the farmers

Hybrids out yield OPVs: Extensive trials have been conducted in the All India Coordinated Pearl Millet Improvement Project that tested hybrids and OPVs in the same experiments. *Hybrids, in general, had about 25% grain yield advantage over the improved OPVs of comparable height and maturity, and 5-6% less dry fodder yield than OPVs* (Table 6.2). The highest-yielding hybrids generally produced 30% more grain yield than the highest-yielding OPVs. Cultivar development research in India got almost all geared to breeding hybrids. There has been downtrend in OPV breeding in India, but OPV breeding continued to be the main strategy in Africa. *At present hybrids cover nearly 50% of the total pearl millet area in India.*

Table 6.2. Grain and dry fodder yield of hybrids and open-pollinated varieties (OPVs) of pearl millet in AICPMIP joint hybrid and population trials in India

Year	Mean grain yield (kg ha^{-1})		Mean dry fodder yield (t ha^{-1})	
	Hybrids[1]	Best hybrid[2]	Hybrids[1]	Best hybrid[2]
1996	1775 (24)	2213 (31)	3.1 (–6)	3.5 (–5)
1997	1676 (10)	2047 (26)	3.4 (3)	4.0 (8)
2000	1117 (26)	1449 (40)	2.0 (5)	2.1 (5)
2001	1917 (27)	2191 (30)	2.9 (–17)	3.6 (3)
Mean	1621 (22)	1975 (32)	2.9 (–4)	3.3 (3)

1. Figures in parentheses indicate superiority (%) over OPVs.
2. Figures in parentheses indicate superiority (%) over best OPV.

Inbred Required for Hybrid Development

The availability and knowledge of cytoplasmic-nuclear male-sterility (CMS), the development of CMS lines, and their maintainers and restorers, made it possible to produce the seed of commercial single-cross F_1 grain hybrids in India. Three types of inbred lines are required for single-cross hybrid production in pearl millet (3 line mechanism) [Inbred line is a nearly homozygous line developed by continued inbreeding (usually selfing) accompanied by selection].

Maintainer line (B line): Line used for maintaining a cytoplasmic male-sterile line. It will have normal cytoplasm and recessive nuclear genes (Fig 6.1). Produces functional pollen grains.

Male-sterile line (A line): A line, in which absence of functional pollen grains is observed in all the individual plants of the line. It will have sterile cytoplasm and recessive nuclear genes.

A and B lines are isogenic lines and usually most of these lines in pearl millet have d_2 dwarfing genes.

Restorer line (R line): Line consisting of all the individual plants with genes (usually dominant) restores the fertility of the male sterile line.

It either has normal or sterile cytoplasm. Restorer line produces functional pollen or male gametes even in the presence of the male-sterile cytoplasm. Most of the R lines used in developing hybrids in India are non-d_2 (tall) genes.

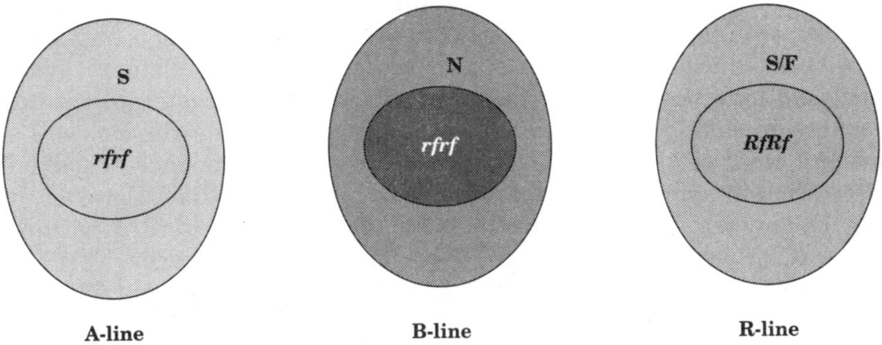

Figure 6.1. Cytoplasmic and genetic constitution of A-, B- and R-lines

A and R lines should be genetically diverse for production of highly heterotic hybrids.

Hybrid Parent Types

Seed parent types

Inbred seed parents: A maintainer line (B-line) can be converted into a male sterile line by repeated back crossing and that serves as new seed parent. A strong B-line breeding is the primary requirement to develop diverse seed parents with potential to produce high-yielding hybrids. Inbred seed parents are commonly used in the hybrid breeding in India and USA.

F_1 seed parents: F_1 seed parents are developed by crossing A-lines with B-lines of unrelated A-line (non-parental B-lines). For example F_1 seed parents developed from a cross between 843A and 862A or ICMA 95111 and ICMB 97444 have been used to produce experimental three-way hybrids.

Advantages:

- Higher seed yield of the F_1s compared to inbred seed parents that is expected to reduce the seed cost of hybrids.
- The dominant nature of the downy mildew resistance can be exploited to increase the longevity of promising but susceptible male-sterile inbred lines through their use in producing the F_1 seed parents.

Except in one case each of three-way hybrid for forage purpose in USA and a hybrid GK 1011 in India, using F_1 seed parents is not a common practice in hybrid development either in India or in USA.

Population seed parents: Converting landrace or improved heterogeneous populations into male-sterile and using it to produce mostly intra-population hybrids. This strategy aims to overcome the problem of adaptation, disease resistance, may be a way of making CMS-based hybrid seed production a more viable proposition in western Africa. The male-sterile population of a Nigerian Dwarf Composite (NCD_2) has been developed, which was chosen based on its downy mildew resistance, grain yield and long panicle traits acceptable to western African agriculture. Conversion of landrace populations into fully male sterile versions was a difficult task with the A_1 CMS system, given the incidence of partial restoration in hybrids made on A-line with unrelated B-line. However, it may be now possible to achieve the acceptable levels of sterility with A_4 and A_5 CMS system, which have higher maintainer frequency. This has been done in two populations (ESRC II and Raj 171). The recurrent selection for male-sterility maintenance for two cycles in ESRC II has increased the male sterile plants from 42% in $81A_4 \times$ ESRC II C_0 to almost 99% in $81A_4 \times$ ESRC II C_2. Similarly, only one generation of selection was sufficient to increase the male–sterile plants in Raj 171 where male-sterility was increased from 42% in $81A_4 \times$ Raj 171 C_0 to 99% in $81A_4 \times$ Raj C_1.

Seed Parent Development

Pearl millet germplasm has a wealth of untapped variability. Evaluation and characterization of pearl millet germplasm has been going on in AICPMIP and core collection of pearl millet germplasm is available. Numerous germplasm sources that can be used to develop superior seed parent progenies include landraces, improved populations and breeding lines, and wild relatives of the primary, secondary and tertiary genepools. The probability of deriving seed parent progenies using wild relatives is low.

Landraces: Landraces are good sources of germplasm for seed parent progeny development if agronomic information is available on individual accessions. Otherwise, identification of needed and/or desirable germplasm from the large number of accessions requires large amount of resources (time, space, monetary and human resource).

Improved populations and breeding lines: Higher probability of developing superior seed parent progenies with new and improved gene combinations could be achieved by using the best sources of germplasm such as: Improved populations, existing inbreds, segregating progenies from crosses between elite germplasm and crosses between elite germplasm and introductions.

Approaches

Pedigree breeding: Selfing within the variable germplasm or in the segregating generations is the most common method in the development of new seed parent progenies.

- Success with this method is enhanced if the progenies are derived form the germplasm or segregating generation with a broad gene base and/or a high concentration of desirable genes.
- The first planting of F_2 populations and composites/OPVs is generally for 500–1000 plants during first year.
- Outstanding populations are replanted in the subsequent seasons to derive more superior progenies.
- Pedigree breeding is generally followed up to F_7/S_5 stage after which bulk breeding is carried out in lines that have attained high levels of uniformity for morphological traits and downy mildew resistance.

Backcross breeding: Back crossing breeding is a popular method used to make a good seed better by introducing one or a few gene(s) from a donor parent. This procedure is most effective with highly heritable traits that can be selected phenotypically. The lines used in backcross breeding will be mostly already established or in the seed production chain or agronomically elite or the one which has proven potential of production of high-yielding hybrid. Hence, during the time of generation advance less selection pressure is required or less testing is necessary once the desired trait is introduced in the new line. However, the danger in using this method could be the narrowing of gene base of seed parents by developing numerous similar elite seed parents each differing only by a few genes. Backcrossing has been highly effective to transfer characters like plant height, earliness and disease resistance that are simply inherited.

Restorer or Pollen Parent Types

The male parent of a hybrid is known as pollinator or restorer line. A male parent may be an inbred line, the F_1 generation of two inbred lines, or a heterogeneous population such as a synthetic or a composite. To date, F_1 pollinators have not been used in pearl millet as inbred lines can produce sufficient pollen.

Inbred pollen parents: Dominant nuclear gene(s) capable of restoring pollen fertility of a CMS line must be present or incorporated into the agronomically superior line that is to be used as the male or pollen parent in seed production of commercial grain hybrid. A large number of inbreds capable of restoring fertility to certain CMS lines, especially those using the A_1 cytoplasm, have been developed in pearl millet.

Top cross pollen parents (TCPs): These are open-pollinated varieties derived from a random mating population that may offer high grain yield. Inbred pollinators are generally considered to be more homogeneous in phenology than topcross pollinators, which may result in nicking problems with changes of day length and temperature of seed production environments. Heterogeneity of a TCP helps to overcome nicking problems to a considerable extent. It has been possible to provide a description of the TCP necessary for ensuring removal of off types during seed production. There are possibilities that TCP can be bred with high yield and greater stability than inbred pollinators for useful traits (*e.g.*, downy mildew resistance and adaptation to heterogeneous environments), while at the same time maintaining relative homogeneity for morphological traits (*e.g*, plant height and panicle characteristics) to facilitate seed certification and gain farmers acceptance.

Restorer Parents Development

Priority CMS systems: Considering the breeding efficiency of the high seed parents with the A_4 and A_5 CMS systems, and their planned use for A-lines development, corresponding emphasis should be placed on breeding restorers of these CMS systems. So far, almost entire emphasis has been on the use of the A_1 CMS system in restorer breeding in all the regional programs in Asia and Africa, except that some of the elite breeding lines have been under conversion into their A_4 and A_5 restorer versions, and A_4 restorers have been introduced into 14 adapted OPVs developed in the ESA region. The low frequency of A_4 restorers observed in several WCA-bred populations point to the need for restorer development research if this CMS system (and also the A_5 CMS system) is to be used in breeding commercial hybrids.

Sources and approaches: Excellent restorers of the A_1 CMS system are available with the public and private sector hybrid programs in India. However, there is a serious lack of the A_4 restorers everywhere. Restorers of A_5 in elite agronomic background are rare and are yet to be developed. Recently excellent genetic stocks of A_4 and A_5 restorers have been developed which can be used for backcross breeding of restorer lines (R-lines) of these two CMS systems. An efficient backcross breeding method (Fig 13.2) for converting elite inbred lines into their A_4 and A_5 restorer versions has been developed. Its application has now led to the conversion of elite inbred lines (designated restorers of the A_1 CMS system) into their A_4 restorer versions (13) and A_5 restorer versions (34), while some more are at various backcross stages of their conversion. Also, moderate to low frequency of restorers of the A_4 and A_5 CMS system has been found in most of the populations surveyed. Their frequency in these populations can be rapidly increased by recurrent selection.

The outstanding R-line materials generated in India provide a good base for their use in pedigree breeding of restorers. Several high-yielding OPVs (and some composites) that have been produced also provide excellent materials for restorer line development.

Majority of R lines is in 46–55 days flowering group, which is now receiving increasing importance in India for breeding dual-purpose hybrids.

Several of these progenies have now been grouped into trait-specific nurseries, following the same philosophy as for seed parent progenies. While these materials will be adequate to serve the medium-term objective, a long-term perspective must include continued trait-based exploitation of germplasm from the genebank until more efficient approaches for the identification of the target traits have been devised and validated.

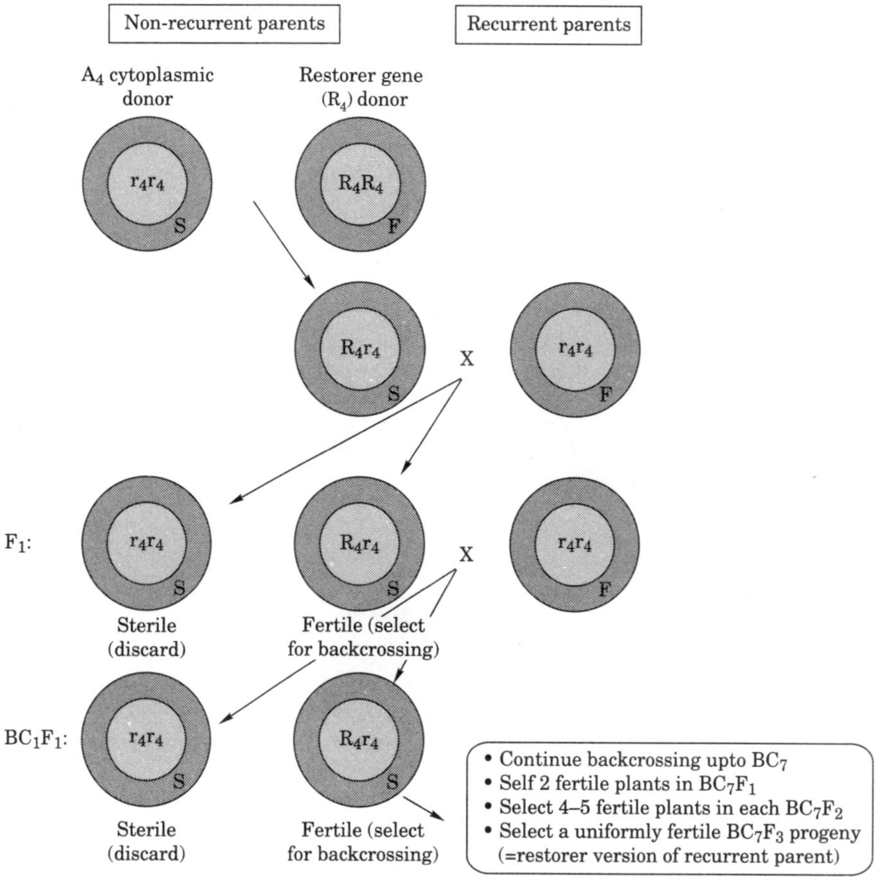

Figure 6.2. Schematic backcross procedure for breeding restorer lines of A_4/A_5 cytoplasmic-nuclear male sterility in pearl millet

Important traits to be considered during selection: Restorer lines must produce profuse pollen that should remain viable at air temperatures as high as 42–44°C. Also, pollen parents must produce highly fertile hybrids, which confers some degree of protection from ergot and smut infection. Besides being able to produce high-yielding hybrids, the restorers should also be highly productive, which is important from the viewpoint of seed production economy. It is desirable to breed pollinators of 150–180 cm height, but no taller than the A-line, with panicle, maturity and tillering attributes that will be preferred by farmers in the hybrids. Pollinators must have acceptable level of resistance to lodging and diseases.

Hybrid Development

Four types of possible hybrids are in pearl millet.

Single-cross hybrids: It is a specific combination of a male-sterile line (seed parent) and an inbred male parent (pollinator) resulting in a high-yielding fertile hybrid (Fig 6.3), the seed of which is commercially multiplied economically. This is the common and most widely cultivated type of hybrid. Important considerations (in order) for the success of this type of hybrid are:

- Yield and downy mildew resistance of the seed parent
- Combining ability of the seed parent.

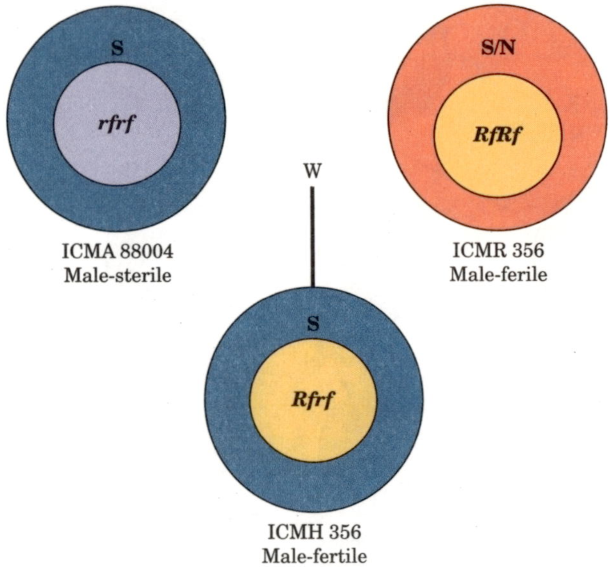

Figure 6.3. Single-cross hybrid production using A- and R- lines, *e.g.*, ICMH 356

Indian pearl millet scenario has changed with the cultivation of this type of hybrid and the productivity gains by cultivating single-cross hybrids are quite significant.

Top cross hybrids: Top cross hybrids (TCH) are bred by crossing an inbred male-sterile line with an open pollinated variety. It may be referred as CMS-line × variety cross. TCHs represent a useful compromise between the higher yield potential of single-cross hybrids and the greater stability in grain yield and downy mildew resistance of OPV. Topcross hybrids could be the best entry point in Africa as the notion of uniform hybrids is not deep rooted in Africa.

Three-way cross hybrids: Three-way cross hybrids (TWHs) are an extension of single cross hybrids that utilize a male-sterile F_1 hybrid (male-sterile line × non-parental B-line) as seed parent instead of an inbred seed parent. The TWHs are likely to yield equally well as single-cross hybrids but may be more stable than single-cross hybrids due to in-built genetic heterogeneity.

Inter-population hybrids: Crossing a male-sterile population with another restorer population produces inter-population hybrids. These hybrid types are aimed at those semi-arid and arid regions of Asia and Africa, where farmers still prefer traditional landraces due to lack of adoptions of single-cross hybrids.

Important consideration for hybrid development: A few important considerations such as flowering time, plant height and panicle types do merit rational judgment while developing hybrids. There should be as less difference between the flowering time of an A-line and R-line as possible. The height of an R-line should not be less than that of an A-line. Having these basic requirements met, hybrid development in pearl millet basically remains a game of numbers. However, identification of heterotic groups in the elite breeding materials can enhance the hybrid breeding efficiency.

Hybrids for various agro-ecological situations: Hybrid development, since beginning has continued to address relatively better-endowed environments. The obvious reasons for this were: (i) greater reliability of results (low trial CV) and fewer crop failures, thus increasing the breeding efficiency; (ii) greater probability of producing high-yielding hybrids, which, by nature, are more responsive and perform better under improved productivity conditions; and (iii) relatively more progressive farmers and farming systems, leading to rapid cultivar adoption. This strategy has paid well and is reflected in rapid hybrid adoption, greater hybrid cultivar diversity, and larger area under hybrid cultivation.

Development of hybrids with high and stable grain yield in more marginal environments, where there is greater poverty and fewer livelihood options, should ideally receive higher priority from socioeconomic point of view. It needs, however, to be kept in mind that development of high-yielding and farmer-adoptable hybrids for such environments is a greater challenge for the reasons just opposite to those mentioned for relatively better-endowed environments. AICPMIP is experimenting through the development of hybrids with available male-sterile lines and restorers and concurrently developing hybrid parents to suite to such conditions

Hybrid programs in India accorded highest priority to grain yield and Downey mildew (DM) resistance combined with maturity duration, mostly in the range of 75–85 days, as per the agro-ecological requirements. Breeding for DM resistance has also been accorded high priority, next to grain yield based on the consideration that there have been several epidemics of this disease in pearl millet hybrids in India as mentioned earlier. Due to the growing importance of pearl millet stover for fodder purposes, there has been considerable emphasis in recent time on breeding for high stover yield. Apart from the evident quality traits both in the grain (size, shape, color) and stover (less rust, thinner stem, lodging resistance) that receive some attention, consideration of any other quality traits in hybrid breeding has been negligible in all the programs.

Alternative hybrid options: Hybrid breeding research in India continues to focus on single-cross hybrids. The three major alternative hybrids include three-way hybrids, topcross hybrids and inter-population hybrids. Considering the stage of hybrid industry development in India, with private sector increasingly playing a dominant role in hybrid breeding and marketing, it is unlikely that these alternative hybrid forms will become commercial in India.

Progress and Achievements in Development of Hybrids

Hybrid development in pearl millet is a long success story. In order to exploit heterosis in pearl millet, attempts were made to breed 'chance hybrid' by planting the parental lines in equal proportion and allowing them to mate in isolation. A few hybrids thus produced were released in the early fifties. They, however, could not become popular because of their limited improvement in productivity over parental lines and lack of organized seed production programmes. Hybrid breeding received a major impetus when cytoplasmic-genetic male sterility was successfully used to develop commercially viable hybrids. India became the first country in the world to develop pearl millet hybrid in the mid-sixties.

The first pearl millet hybrid HB 1 was released in 1965, followed by millet growing areas and helped to raise the production level appreciably in a short span of 10 years. Intensive cultivation of hybrids in the early 1970's led to downy mildew epidemic. Therefore, a greater emphasis was given after 1975 on development of hybrids that combined resistance to downy mildew disease with high production potential. The new hybrids like BJ 104, BK 560 and MBH 110 developed in the 1980's were widely grown. Specific efforts of the past decade have also resulted in the release of very early maturing and drought tolerant hybrids, viz., HHB 67, HHB 68, ICMH 356 and GHB 15 for various pearl millet growing regions. These hybrids, in the maturity range of 60–70 days, have the yield potential of 22–25 q/ha. Recently, some more promising hybrids combining high level of grain yield and downy mildew resistance are released and notified at national/state levels. They include GHB 577, RHB 127, HHB 117, HHB

146, GHB 558, GHB 526 GHB 538, HHB 67-2, GHB 715, GHB 719, GHB 757, GHB 744, GHB 732 and HHB 197. The hybrids released during the last decade are based on a large number of genetically diverse male-sterile lines and pollinators with the result that downy mildew epidemics are largely under control.

Maintenance and Production of Parental Lines and Hybrids

Seed class: Hybrid is the commercial end products in pearl millet which is made available to farmers. Once the decision is taken to release a hybrid with specific characteristics, the pearl millet breeders make a limited increase of seed of new hybrid parents. Seed multiplication of any class depends on the actual requirements and the buffer stock. In pearl millet, seed is multiplied in four stages (nucleus, breeder, foundation and certified), called seed classes.

Nucleus seed: It is the first stage in the chain of seed multiplication of or hybrid parents. This fundamental stock of seed obtained from the selected individual plants of parental lines of a hybrid (A, B, R line) is the only seed that can be used to produce its own seed class. Nucleus seed should be of the highest genetic purity and the responsibility for seed production lies with the originating breeder.

Breeder seed: of hybrid parents is produced by or under the direct control of the sponsoring plant breeder. This class of seed is the base of the first and recurring increase of foundation seed and is required to be inspected by a monitoring team consisting of a breeder, seed certification officer, representatives of National Seed Corporation (NSC) and/or State Seed Corporation (SSC).

Foundation seed: of hybrid parents are produced from breeder seed, the production of which is carefully supervised or approved by the breeder and seed certification agency at the experimental station or recognized seed farms. It is a source for producing certified seed class and should satisfy the minimum seed certification standards.

Certified seed: of a released hybrid is produced by registered seed producers, duly certified by seed certification agency and produced on a large scale for general cultivation. The linkage between seed classes is given in Fig 6.4.

Guidelines for Seed Production

Nucleus seed production: Nucleus seed of parental lines refers to the seed produced by breeder who developed the particular hybrid or by any other breeder of the institution where the hybrid was developed which is directly used for multiplication of breeder seed.

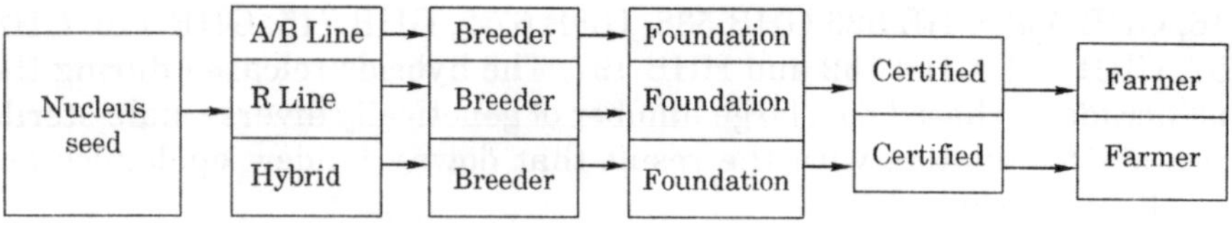

Figure 6.4. Linkage between seed classes

Nucleus seed is the first stage in the chain of seed multiplication of hybrid parents and is the only seed that can be used to produce its own seed class. To start nucleus seed programme of parental lines, the basic source of seed (seed maintained by breeder) is a prerequisite. It is advisable to produce nucleus seed during off-season or under strict selfing to ensure genetic purity and homogeneity. The purity should be for all the characters associated with the identity of a variety or line. The procedure for the production of nucleus seed of male sterile line, maintainer line (Fig 6.5) and restorer line in pearl millet is described below.

Hybrid parents: The nucleus seed production of hybrid parents essentially involves the seed production of A line (male sterile), B line (maintainer) and R line (restorer) as discussed below.

Maintainer Line (B lines)

Season I:

- Grow large number of plants of B line (0.05 ha)
- Select and self about 1000 plants at the time of flowering
- Finally select about 200 selfed plants conforming to the characters of maintainer line.

Season II:

- Grow plant to row progenies of selfed plants in two replications retaining the remnant seed
- The progeny rows are studied for the diagnostic characters and rows not confirming to the characteristics of the line are rejected
- Identify the best progeny rows (25–30%)
- Bulk the remnant seed of selected best lines.

Season III:

- Grow the bulk of remnant seed in isolation
- Bulk the seed of all the plants after the harvest
- This forms the nucleus seed bulk of B line

(ii) MS line (A lines):

Season IV:

- Grow A and B lines in alternate rows. Seeds of B line will be those obtained from rejuvenation as given in the maintenance of B lines
- Make 200–250 paired crosses between A and B plants
- Care must be taken to cross A and B plants confirming to the line standards only
- Paired crosses among A and B lines should be labeled viz. $A_1 \times B_1$, $A_2 \times B_2$ etc. and harvested seed of each pair should be kept separately

Season V:

- Grow respective paired crosses involving A line (crossed seed) and B line (selfed seed) in alternate rows
- Retain a portion of seed as remnant seed
- Observe critically pairs of A and B lines for all the characters including height, flowering and typical morphological characters
- Observe for pollen shedders in A lines. The A line progenies showing pollen shedders and corresponding B lines should also be rejected
- Identify uniform pairs of A and B lines which conform to the standards of parental lines
- Remnant seed of the A lines of the selected pure pairs is bulked.

This forms nucleus seed bulk of A line.

(iii) Restorer line (R lines)

Season I:

- Grow large number of plants of R line (0.05 ha).
- Self a number of plants (about 1000) conforming to the standards of the line at the time of flowering.
- Finally select about 200 selfed plants based on field studies as well as observation in laboratory for seed colour, shape etc.

Season II:

- Grow plant to row progenies of selfed plants in two replications.
- Retain a portion of selfed seed of each plant as remnant seed.
- The progeny rows are studied for the qualitative (diagnostic) characters.
- The progenies not conforming to the characteristics of the line are rejected.

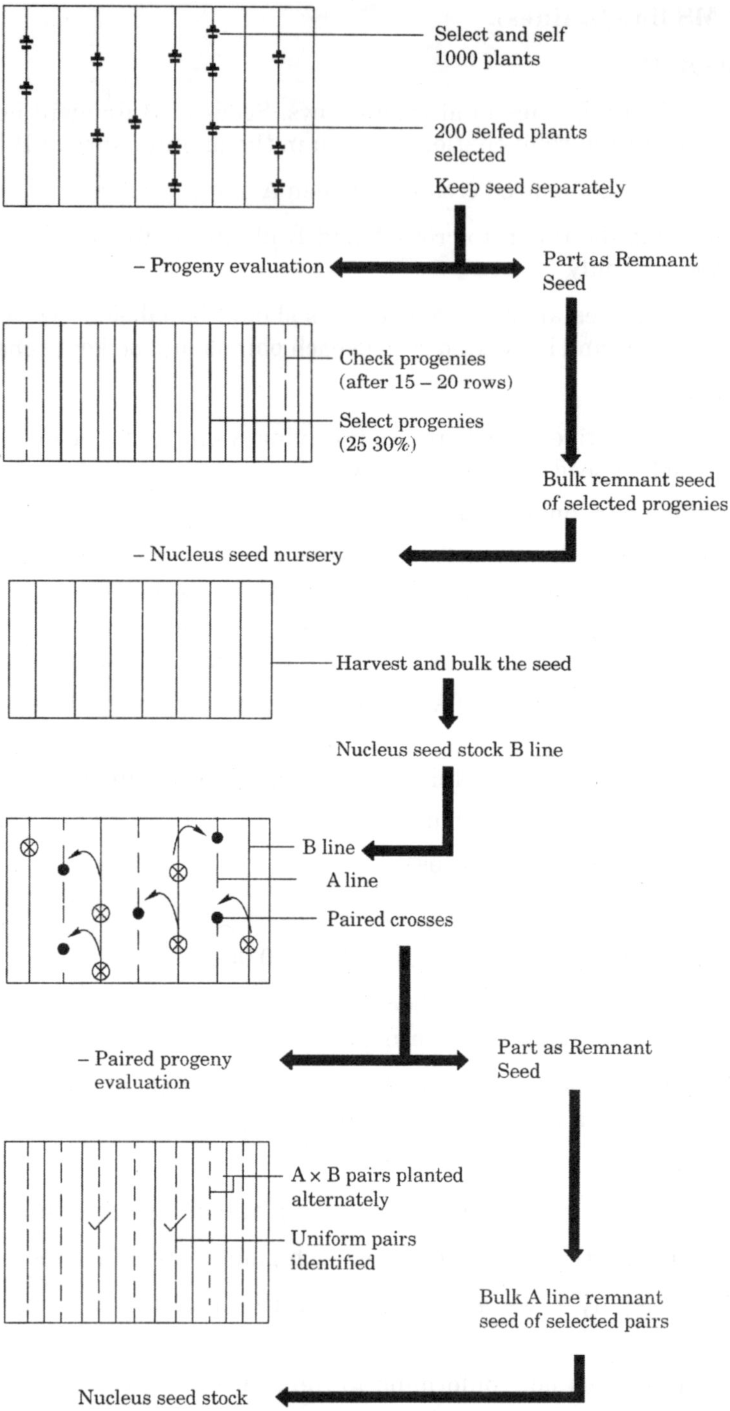

Figure 6.5. Nucleus seed production of maintainer (B) line and male sterile (A) line

- Evaluate the lines for yield and agronomic score and for other economic characters.
- Identify the best progeny rows (30 to 50%) based on all characteristics mentioned above. In the progeny row testing if adequate number of progeny rows conforming to the line standards are not obtained, selfing for one or more generations will be required. These selfed plants of R line should also be tested for their restoration ability.
- Bulk the remnant seed of best lines.

Season III:

- Grow the seed of remnant seed in isolation.
- Bulk the seed of all the plants after the harvest.
- This forms the nucleus seed bulk of R line.

Breeder/Foundation Seed Production

Breeder seed is produced from nucleus seed stock while foundation seed is produced from breeder seed as per detail given in seed class. The minimum isolation distance is 1000 m for breeder/foundation seed of A, B, R lines and OPVs.

Male Sterile (A) Line and Maintainer (B) Line

- Multiply A line by planting A and B lines in alternate sets usually in the ratio of 4:2.
- Plant 4–8 border rows of B line around the production block to ensure adequate pollen supply to A line.
- Careful and strict rouging is a necessity in A and B lines. Rogue out pollen shedders in the A lines, if any.
- Harvest B-line rows immediately after completion of flowering period.
- Carefully harvest the A line rows and bulk the seed.
- For breeder seed production of B lines raise separate production plot of 0.1 ha area under isolation (1000 m) as bulk planting from nucleus seed stock.
- Follow rouging and other procedures as detailed in nucleus seed production method.

Restorer (R) Line

- Breeder/foundation seed production of R lines is done by bulk planting under isolation from the nucleus, breeder, and seed stock.

- Seed plots should be 0.1–0.2 ha for R lines and at least 3000–5000 plants should be maintained.
- Crop is inspected at the following different stages of growth to identify deviants for different characters:

Stage of crop	**Characters specifically observed**
18 days after sowing	Identify deviants for early vigor
30–35 days after sowing	Identify deviants for tillering, plant height and other traits
Boot leaf stage	Identify deviants based on early flowering
50% flowering	Identify deviants for late flowering, panicle exertion and panicle type
Maturity stage	Identify deviants for seed type and grain yield

- The identified off-type plants should be removed by uprooting as cutting of plants may lead to ratoon tillers.
- Select 500–1000 panicles (depending on the requirement) from the bulk planting.
- Harvest and bulk the seed.
- After harvest, the seed should be dried, processed and stored in safe containers.

Certified Seed

Certified seed of hybrid is produced in pearl millet for farmers' use. The stipulated isolation distance for certified seed production, for hybrids is 200 m. The pattern of planting and seed production is the same as for the breeder and foundation seed production plan. The production must be acceptable to seed certifying agency and fulfill all requirements of certification.

Certified seed is generally produced through contract growers. Some private seed companies also undertake certified seed production programmes.

The certified seed of a hybrid is produced by growing male sterile line with a specified restorer line in an isolated field (Figs 6.6 and 6.7). Synchronisation of A and R lines is crucial for certified hybrid seed production. This can be manipulated by (1) differential dates of sowing (2) manipulating of water and fertilizer to one of the hybrid parents, and (3) removal of extra early tillers in A or R line to synchronise the ability of pollen shedding and stigma receptivity.

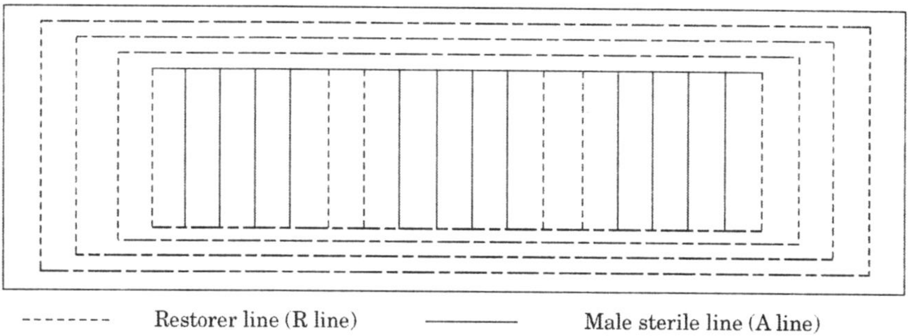

Figure 6.6. Showing pattern for certified hybrid seed production in pearl millet

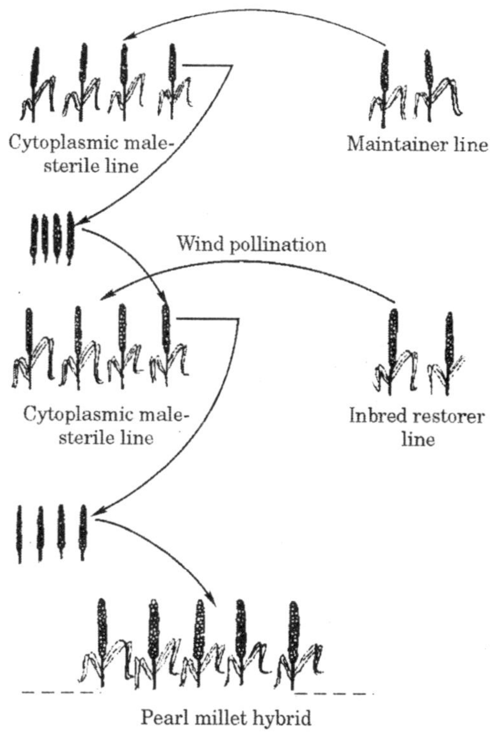

Figure 6.7. Scheme for pearl millet hybrid seed production in an isolated field

Off-type plants in the R line and pollen shedders in A line should be roughed out carefully to maintain genetic purity. A satisfactory certified seed production can be achieved if seed village concept is followed.

Seed Production Management

In order to obtain high quality seed with good physical appearance, uniformity, large grain size and high germination per cent, proper agronomical practices should be adopted as detailed below.

The seed production plots require extra care. The area chosen should be uniform, well levelled, free from disease inoculum, weeds and cross compatible wild species of pearl millet. Avoid areas where rain occurs during seed development stages. A few rains, or even continuous high humidity during seed ripening period may cause mould development, or even seed germinate on panicles, discolouring them and substantially reducing germination and storage life. The land on which pearl millet crop was grown in the previous season should be avoided to minimize the problems of volunteers.

The land should be well prepared (15–20 cm deep) for seed production plots. It is advisable to plant seed on ridges at a distance of 45 to 60 cm. Planting is done by drilling, dibbling or transplanting. Drilling places the seed at a uniform distance and depth, enabling uniform plant stand. Further, large areas can be quickly planted by using a precession fertilizers-cum-seed drill mounted on a tractor or drawn by bullocks. Dibbling by hand is a common practice where labour is cheap and fields are small. Transplanting is done practically in all the southern states to save valuable seed material. About 0.03-hectare nursery is enough to provide seedlings for one hectare. It reduces the life span of the field crop by 18–21 days. Transplanting also enables easy adjustment of staggered plantings.

To ensure optimum plant populations (120,000 plants) in a hectare, with 45 cm distance between rows and 12 cm between plants, 4 kg seed is required for the multiplication of restorer lines and OPVs while 3 kg of A line and 1 kg of B line or R line is needed for the multiplication of A line or hybrid, respectively.

Manure and fertilizer requirements include 8–10 tons of farmyard manure or compost, 100 kg N, 60 kg P_2O_5 and 40 kg K_2O per hectare. Fertilizer doses may split into two; a basal dose of 40 kg N and all P_2O_5 and K_2O at the time of planting and topdressing of nitrogen in two equal splits at tillering and panicle emergence may be practicised.

Water is an essential input to ensure good seed yield. The most critical stages to irrigate pearl millet seed crop are tillering, flowering and seed development. Moisture stress at any of these stages reduces seed yield considerably. Efficient drainage in seed field is also essential as pearl millet is susceptible to water logging particularly in the early stages of crop growth.

Weeds claim major share of soil moisture and nutrients, and reduce yield of seed crop. They may act as alternate hosts for various diseases and pests, and also create problems during harvesting and threshing. Weeds can be efficiently checked through two interculture operations at an interval of 20 days in the early stage of crop growth. Use of herbicides greatly reduces losses due to weed competition.

Pearl millet is a high value seed crop. It is therefore, essential to control disease and pests, which are likely to reduce the seed yield and quality. Cultural control methods like clean cultivation, planting date adjustment, and uprooting of disease-affected plants can substantially reduce the damage and minimize seed yield losses. Some times, diseases and pests attack may be controlled through chemicals.

In a seed crop, harvesting may influence the physical quality of seed, processing requirements, storability and seedling vigour. If the panicles are artificially dried and threshed mechanically, then harvesting can safely be done when the seed attaines physiological maturity (25% moisture content).

The panicles should be dried to 12% moisture content. Proper care must be taken in harvesting particularly of hybrids and A/B lines multiplication to avoid mechanical mixture from lodged plants. Care must also be taken during threshing operations. Proper handling of all class of seed in post-harvest operations should be of great help not only to maintain seed quality but also to prevent post-harvest losses in seed quality. Threshing of nucleus/ breeder seed should be done by small size seed thresher. These threshers should be thoroughly cleaned. After threshing, seed should be dried and kept in sealed containers.

Seed Processing

Pearl millet seed crop may be harvested when moisture in the seed is in the range of 19–22 per cent. Seed lots received from the field often have high moisture content and contain trash and other inert material, weed seeds, deteriorated and damaged seeds etc. Seed processing is a vital part responsible for upgrading the quality of the seed by removing foreign material and undesirable seeds.

Drying: Moisture content is an important factor for seed quality and considered prior to processing. Seeds with moisture content above 15 per cent are subjected to excessive damage in processing. Hence natural or artificial drying is necessary. Sun drying is mostly practised in pearl millet to bring down moisture to the safe processing level of 12% or less. Solar-cum-agriculture waste fire dryers can be used for pearl millet during inclement weather. These dryers are available with a capacity of 800 kg seed. The cost of a dryer is approximately Rs 10,000 and drying cost of one quintal seed is about Rs 2/- only.

Seed cleaning and upgrading: Seed cleaning and upgrading mainly depend on seed size (length, width, thickness), density, shape, surface texture, colour etc. If the differences between desirable and undesirable material with regard to any of these properties exist, separation of undesirable material should be done with the help of suitable machine.

Pearl millet seeds are either round or oblong. Screens for cleaning are selected according to the shape of the seed. A top screen with round holes or slotted screen may be used for round seeds, while for oblong seeds, top and bottom screen with oblong holes may be used. The suggested screens for seed cleaning in pearl millet are given in Table 6.3.

Table 6.3. Suggested screens for seed cleaning in pearl millet

Cleaner	Place of screen	Number and size of screen
Four screen	Top screen upper shoe	9,8
	Bottom screen upper shoe	14 × 14, 1/14
	Top screen lower shoe	8, 1/12 × 1/2
	Bottom screen lower shoe	3 ×16 Sp., 3 × 14
Three screen	Top screen	9,8
	Middle screen	8, 1/12 × 1/2
	Bottom screen	3 × 16 Sp., 3 × 14

*A. T. Ferrel and Co.

In certain cases, it may be necessary and in others rather desirable to further upgrade the seed quality by removing either specific contaminants, or using very precise size upgrading operations. The choice of upgrading however shall depend on the type of contaminants. Various types of upgrading equipments are utilized for this purpose.

Seed treatment: Seed treatment refers to an application of fungicides, insecticides or a combination of both to seeds. The purpose of seed treatment is to disinfect and to protect the seed from seed borne, soil borne pathogenic organisms and storage insects. Pearl millet seed is also protected against downy mildew, ergot and smut diseases.

Seed packing and labeling: Seeds are packed in containers of specified weight after processing and treatment. Cloth bags of 3.0 or 1.5 kg capacity are utilized in pearl millet. Cloth bags of 5.0 or 8.0 kg capacity are also used. A label must be placed in or on the bags to maintain positive identity of the seed. The outer label or tag can be sewn to the bag when bags are closed with a sewing machine. The label must carry the following information:

Crop ______________	Label No. ______________
Variety ______________	Pure seed (%)* ______________
Class of seed ______________	Inert matter (%)* ______________
Production institute ______	Germination (%)* ______________
(Name, address and seal)	Genetic purity* ______________
	Date of test ______________
	Net contests ______________

(Packed at __________% moisture)

*Based on actual

Seed storage: Good seed storage is a basic requirement to maintain the quality of seed. Seeds are stored either in field after harvest, until processed and taken to seed stores, in seed stores, in transit (carts, trucks, railway sheds, rail wagons etc.), in retail store, in the farmers house, or on the farm. Germination and vigour may be considerably affected in the above storage situations. So, proper care should be taken while storing seed at different places. The basic steps in seed movement through a processing plant are divided into three phases as illustrated in Figs 6.7 and 6.8.

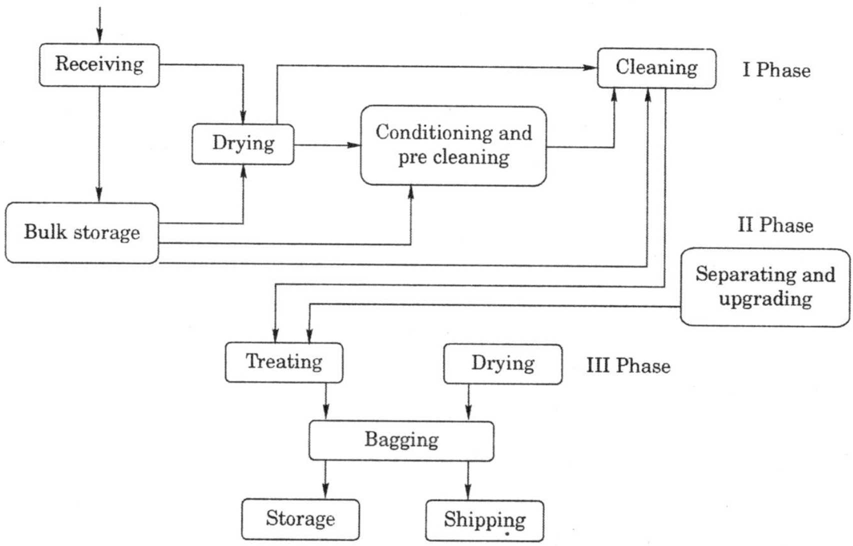

Figure 6.8. Basic steps in seed movement through a processing plant

Seed Marketing

Pearl millet hybrid seed requirement: Assuming 50 per cent hybrid coverage, India requires 22,500 tonnes of certified seed of pearl millet and hence pearl millet hybrid seed market is estimated around Rs 15,00,000.

Seed sales: The National Seeds Corporation (NSC) and State Seed Corporations (SSCs) are the public sector organizations chiefly involved in the distribution and marketing of seeds. NSC seeds reach the farmers at the state, district, tehsil and village level through distributors or through their sale counters at the tehsil level. This prevents the involvement of middle man and ensures check on prices and quality. The SSC seed is supplied through appointed distributor, state agricultural department seed stores or through SSC sale counters. The NSC and SSCs also distribute seed *via* agro industries corporation and government godowns (Fig 6.9).

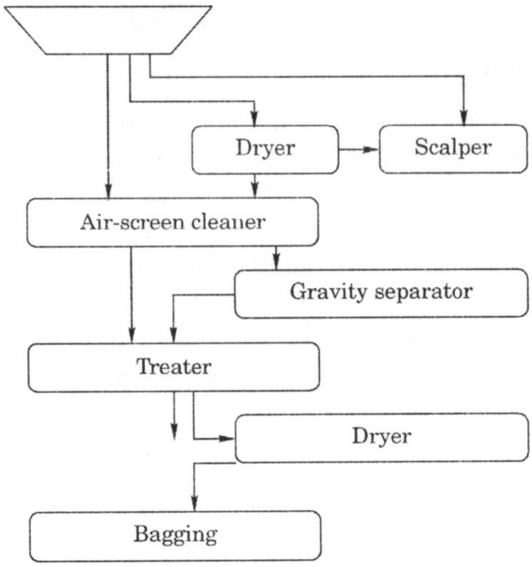

Figure 6.9. Flow chart of Pearl millet seed sales

The private seed companies follow their own marketing and distribution channels. Their seeds reach farmers *via* one or more distributors at district or town level. Each distributor has a number of dealers at the village level, who sell seeds directly to farmers (Fig 6.9). In general, dealers are responsible to the company's distributor, and sell seed on prices fixed by the company. The distributors receive 12–15 per cent discount or commission from the company, who passes on 10% to the dealer. In many areas, the company may often sell seed direct to the farmers through retail depots.

The seed pricing policy is the most effective but delicate issue for regulating the flow of seed according to the plans. The price fixed for sale should be responsible to attract a large number of farmers to purchase the certified seed. However, prices of certified seed are rather high since they include the cost of production, processing, certification, storage, marketing, and interest on capital, transportation and dealer's commission.

Private seed companies' prices are 50–100% higher than the public sector agencies because they also include the research costs. Generally differences between retail price of public and private seed of a hybrid is considerable. It indicates that private pearl millet hybrid seed on an average are more expensive than public bred hybrid. There is no standard formula for calculating the cost of certified, foundation or breeder seed. Foundation and breeder seed prices are related to price of certified seed. The seed companies in the public and private sector are free to fix price for their own seed. Normally, the cost of certified seed is two to three times more than that of procurement/spot price of the commercial grain. The price of the breeder seed is usually 50% higher than that of foundation seed.

Generally farmers purchase certified seeds on cash basis. The National Seeds Corporation (NSC) and State Seed Corporations (SSCs) do not directly give credit to farmers for purchase of seeds. However, credit can be obtained from the cooperative societies. Often poor farmers fail to purchase the costly seeds in the absence of this facility. However, some private seed companies sometimes offer 10–45 days credit through dealers depending on the credit worthiness of the buyer and the farmers' relationship with the trader.

Popularization of Hybrids

Hybrid pearl millet sales would increase in coming times with more replacement of the OPVs as well as by beating the competition due to strong research. The semi-dwarf segment of Maharashtra would significantly increase due to shift from tall OPVs to hybrids. The large area in Rajasthan would also have more penetration of tall hybrids in coming times.

The evolution of new pathotypes of the downy mildew pathogen is a constant threat to the existing products in all market segments. This problem needs regular attention to give resistant hybrids to the upcoming pathotypes regularly.

The critical success factors in the pearl millet seed business are as follows:

- Suitable product for the concerned segment
- Quick replacement of old products with good promotion efforts
- Good quality control of commercial hybrid seed
- Availability of hybrid seed in the market ahead of rains because major area is rain dependent and the sales window for sales is short.

The marketing and sales success depends on proper planning much ahead of the season and its successful execution. Planning involves understanding the needs of the farmers based on the rainfall and soil type, activities to be undertaken by the M&S team, feedback etc. The activities for M&S of pearl millet hybrids are as follows:

Commercial Activities

(i) Advance booking scheme (ABS)

(ii) Motivation channels:

  (a) Distributor conference

  (b) Retailer meetings

(iii) Stock placement

(iv) Commercial schemes:

(a) EPD

(b) TOD

(c) Retailer schemes

(d) Target linked trips

Demand Generation Activities

In the markets already covered under hybrids

(a) Through innovative sales promotional farmer scheme.

(b) **Films on farmer scheme:** It is the marketing of marketing tools by making a film regarding innovative farmer schemes, which may be shown at village level through video on the wheels.

(c) **Training the field assistants:** To train the field assistants about the key benefits of the product, to help him spread the key message regarding the farmer schemes and the product.

(d) **Rickshaw campaign:** Manual rickshaw along with the sun packs, loud speakers, with an audio cassette as a mass communication at the village.

(e) **Hoardings:** 9"–12" flexo hoardings at the mega distributor shop or at renowned public places to popularize the farmer scheme.

(f) **Millet suggestion centre:** In key markets along with small tent, the staff provides suggestions to farmers about millet cultivation technology.

(g) **Chetana rath yaatra/ video on the wheels:** Innovative video on the wheels along with decorative flexo of farmer scheme and product visuals.

(h) **Big farmer meeting:** A planned formal farmer meeting of more than 200 farmers which may be interactive and innovative.

(i) **Mega farmer meeting:** A planned formal farmer meeting of more than 5000 farmers which is the most effective activity to create the hypergarding the product as well as the scheme.

Development of New Markets

This change can be achieved by activities like *Parivarthan rath yaatra*. As the name indicates, *Parivarthan* means conversion from open pollinated varieties to hybrids. A decorated and attention-getter video on wheels may be used to convince the farmer about the benefits of the hybrid over OPV varieties along with crop economics chart as follows:

Based on the efforts made by the sales and marketing departments of major seed companies, the change from OPVs to hybrids is being observed. This effort would be faster if the new hybrids are developed keeping in view the needs of the farmers growing these OPVs.

The sales and marketing activity is thus very important to carry the fruits of the R&D activities to the end users, *i.e.* the farmers. Pearl millet market needs to further strengthen these activities.

Identification of Alternative Locations

The seed production of most pearl millet hybrids is generally undertaken in southern parts of the country because of congenial agro-climatic conditions and possibility of taking two seed crops in a year, if necessary. The prominent states are Gujarat, Maharashtra, Andhara Pradesh, Karnataka and Tamil Nadu. In contrast, the northern and eastern states from Punjab to West Bengal dominate rice, wheat and legume crops seed production which is less profitable. Consequently there is a major difference in the cadre of seed industry between the North and West/South of India. Technology of seed production is moving at a rapid pace. The companies ensure that they expeditiously pass on the new technology to their contract seed multipliers to harvest high yields of good quality seeds. Most public bred hybrids and variety seeds are certified while the quality monitoring of proprietary hybrid seeds of private sector is through their in-house quality control staff and the seeds are sold under truthful label. All seeds meet minimum prescribed standards. Opportunities are also available to produce quality seed in north Indian states like Rajasthan, Harayana, U.P., and Gujarat. These possibility need to be exploited in pearl millet seed production to reduce transportation costs.

Linkages between Public Private Sectors

Till such time the population stabilizes, the demand for food would continue to increase. To keep pace with the present rate of population growth and consumption pattern, India's food requirement has been estimated to cross 300 million tons by 2020. This would mean an annual agricultural growth rate of over 6 per cent (from current 2.2 per cent), a daunting task considering the rapidly shrinking resource base and vast declining input use efficiency in major cropping systems.

To supplement public research, private R&D would need to play a greater role in finished product breeding and generation of relevant technologies. It should strengthen its research infrastructure to play a crucial role in evolving, identifying and introducing superior value added hybrids with wide and/or area specific adaptability.

The later will be particularly relevant for specific soils or agro-climatic conditions of small and marginal farmers. This is perhaps an appropriate time to restructure national crop breeding approach again wherein private sector plays dominant role in:

- Availability of superior value added products suitable for diverse agro-climatic and soil variations.
- Breeding adaptable value added area specific for all groups of farmers.
- Developing applied production technologies for favourable, as well as, hostile (rainfed) crop growing conditions.
- Improved product quality.
- Follow up effective extension, to ensure that superior hybrids and technologies developed in research laboratories expeditiously reach farmers' field.

NARs and IARCs major focus should shift to breeding new parental lines, germplasm augmentation, identifying new sources of resistance, new gene constructs, biotic and abiotic stress and integrated crop production technologies and making them available to private sector on payment. This may be done in a phased manner. Private sector should breed value added end products, produce their high quality seed for all production environments and aggressively market them.

Technology will drive future growth in agriculture, through judicious use of molecular biology, biotechnology including genetic engineering and other new cutting edge technologies, to add value through seed in the products being commercialized. It is now clear that effective technology transfer, involving a continuance of technology generation, assessment, refinement and transfer is needed to realize the gains at the farmer's fields. Testing, multiplication and supply of improved value added seed is, thus, warranted where private sector plays a greater role. A synergy of both private and public sectors will not only be complementary to each other, but will also take the modern technology to the door-steps of the farmers at an affordable price and much faster pace. Agricultural biotechnology possesses enormous potential. Used in consort with conventional breeding approach and harnessed judiciously, it can help raise crop productivity, introduce resistance to pests and diseases, heighten tolerance to adverse weather conditions, improve the nutritional value of some foods and enhance the durability of products during harvesting or shipping. Agricultural biotechnology therefore can directly contribute to feed many more people in a sustainable way and bring the rural poor out of poverty.

Indian seed industry has reached a stage of maturity and is in a position to shape itself to compete at national and international levels. It may, however, be recognized that private sector for some more time will continue to concentrate on more lucrative sector of the market, with a strong emphasis on hybrids and the more specialized low volume crops. Fundamental strength and potential of Indian seed market cannot be doubted. The country has a very large population and a strong commitment to self-sufficiency in food production, which can only be achieved by increasing productivity per unit area.

Indian seed industry now needs to become technology driven and innovation led. The new technologies and value added seeds through appropriate GOI policies can be made accessible even to resource poor marginal farmers. This transition will have to be facilitated through strategies where both the government and industry are partners. Agriculture can be best served when the right mix of technology, ingenuity and business savvy is brought together, with active government support. Seed industry worldwide is going through a rapid technology revolution with value added output traits at its center, driven by consumer and commercial end users. It involves all aspects of agri-biotechnology and crop production. Emphasis is on proprietary genes in value added crops for the feed, food and chemical industries. This change to value added agriculture will also have major impact on restructuring of Indian seed sector, hence seed business in the coming decades. In the changing seed industry scenario, the industry would need to become technology driven and innovation led. This transition will have to be facilitated through innovative strategies where both government and industry are partners.

Challenges posed by the globalization of agriculture and government's progressive economic reform agenda provide excellent growth opportunities in the seed sector. Indian seed industry can look forward to producing and marketing "paleted seed capsules" designed for specific situations with high inherent genetic yield potential, genetically engineered or chemically induced tolerance/resistance to major pests and/or diseases, drought tolerance or tolerance to salinity, attractive appearance, higher nutritional content, longer keeping quality, or desired end-use characteristics resulting in plants with less dependence on fertilizers or crop protection chemicals and above all provide stability of production under favourable as well as unfavourable environments. A large cadre of highly skilled seed multipliers exists. Companies both in public and private sector have developed systems for training technical manpower. As their R&D gains strength, they will develop new hybrids suitable not only for Indian market but also for the agro-climatic conditions prevailing in the export market. India's low cost skills at all levels of seed industry, from research through seed production, coupled with range of environments, suitable for production of most crop seeds and rapid flow of technology, offers excellent base for competitive seed supplies for domestic and international markets. High quality improved variety seed has already been accepted by the educated, well-to-do farmers. They have derived benefits by using superior genetic material and the associated technology by harvesting high yields. They realize that seed is the cheapest input in their total cost of cultivation and hence are willing to pay its fair price. They have accepted high-tech value added seed and are convinced of their benefits. The challenge in the coming years would be to convince the small and marginal farmers to accept quality seed of recommended, adaptable and value added improved hybrids. Regional private seed companies would probably specialize in specific farming systems, to cater to the needs of such farming families in their entirety by developing area- specific technology, considering the economic condition and the agro-climate of the region. These companies would play a crucial role as catalyst in transferring such technologies from research laboratories to the farmers' fields.

Indian economy is opening up the opportunities and challenges offered by the globalization of agriculture, particularly in the seed sector. There is a need for a long term "Seed Policy" to recognize and reward innovations, replace multifarious laws governing seed through one comprehensive legislation, promote creation of high-tech R &D, production and conditioning facilities to produce value added seed, meeting international standards resulting in plants with less dependence on fertilizer or crop protection chemicals and ensuring stability of production in favourable, as well as unfavourable environment. Indian seed industry is on the threshold of becoming a global player. Indian private sector can look forward to playing a key role in the development of Indian seed industry and take its rightful place in the global market.

Protection of Hybrids

At the highest level government policy is to reduce controls and to make the economic system more transparent. In practice, however, there is potential proliferation of legislation affecting seed sector. Presently the Indian seed industry is governed by the Seed Act 1966, Seed Rules 1968, Seed Control Order 1983, Essential Commodities Act 1955, Package Commodities Order 1975, Standards of Weights and Measures Act 1976, Consumer Protection Act 1986, Plants, Fruits and Seeds (Regulation of import into India) Order 1989, Urban Land Ceiling Act 1976, State Acts of land acquisition/land use and State Acts for the control/movement of crops and seeds.

The policy framework hitherto has been regulatory with excessive legislation. The above plethora of legislation covering the seed sector is out of date and restrictive rather than facilitative. The multiplicity of Acts, rules and administrative orders have resulted in overregulation of the nascent industry. One comprehensive seed act should replace all the above acts.

The Protection of Plant Varieties and Farmer's Rights Bill (PPVFR) has been passed. PPVFR authority has been established at IARI campus at Pusa, New Delhi in 2006. Some of the centres for DUS testing have been provisionally identified. Salient features of PPVFR are:

- All plant varieties to be registered.
- Farmer groups to be rewarded for contribution in evolution of varieties.
- Farmers can save seed for own use or barter/exchange with neighbours.
- Allows free and complete access for researchers.

The revised seeds act has been referred to the "Select Parliamentary Committee" by the union cabinet. Registration of all cultivars would become mandatory when this act is implemented. The registration will be based on "Value for cultivation" tests and this will protect public bred varieties and hybrids.

Benefit-Cost Ratio

The benefit of hybrid cultivation is approximately two and half times greater than by growing *desi* or open-pollinated varieties, for example

Input (Rs)

S. No.	Operations (inputs)	*Desi* (OPV) (Rs)	Hybrid (Rs)
1.	Land preparation and sowing	150	150
2.	Manures	300	300
3.	Fertilizer	500	500
4.	Irrigation with labour charge	250	250
5.	Manual weeding (2 labourers)	200	200
6.	Harvesting (3 labourers)	300	300
7.	Seed cost	80	225
	Total	**1780**	**1925**

Output

	Desi (Rs)	Hybrid (Rs)
Grain yield	3200	6000
Dry fodder yield	1000	1600
Total	4200	7600

Profit from *desi*/OP variety per acre: 4200 – 1780 = Rs 2420

Profit from hybrid per acre: 7600 – 1925 = Rs 5675

Difference = 5675 – 2420 = Rs 3255

Extra investment = 145

Extra income = 3255 Net Profit = Rs 3110

7 Strategies for Commercialization of Public - Bred Hybrids of Castor, Sunflower and Safflower in India

A.J. PRABAKARAN AND D.M. HEGDE*

ABSTRACT

Sunflower, castor and safflower are important oilseed crops where heterosis has been exploited extensively on commercial scale by both by public research organizations and private sector companies utilizing the cost effective cytoplasmic male sterility-fertility restoration system in case of sunflower, genetic male sterility in safflower and unique pistillate lines in castor. The availability of adequate quantity of quality seed is a major limiting factor for the spread of hybrids. Flowering habit, male sterility system, commercial production of hybrid seed, maintenance of parental lines, achievements and strategies for popularization of hybrids are presented.

Key words: Castor, heterotic potential, parental lines, safflower, strategies, sunflower

The demand for vegetable oils is ever increasing despite impressive gains being achieved in oilseeds production in the country to record new landmarks, forcing the country to resort to huge imports. Among annual oilseed crops, area and production of safflower and castor in India dominates the world while sunflower has bright prospects.

Table 7.1. India's position in world's oilseeds (1999–2006)

Oilseed	Area ('000 ha)		% Share	Production ('000 Mt)		% Share
	World	India		World	India	
Sunflower	21642	2050	9.5	27979	1230.0	4.4
Safflower	962	380	39.5	738	216.0	29.3
Castor1388	809	58.3	1330.0	869	65.3	

Sunflower, safflower and castor currently contribute about 12% of oilseeds area and 10% of oilseeds production of the country. Nevertheless,

* Directorate of Oilseeds Research, Hyderabad 500 030.

these crops constitute a pride of place in the basket of vegetable oilseeds in terms of their high growth rates (sunflower), high quality of edible oil (safflower and sunflower) and unique industrial use (castor). The adaptability of sunflower to wide agro-climatic conditions, the drought hardiness of safflower and castor, increasing demand and market price for these crops make them competitive choice for inclusion in crop diversification.

Sunflower

Sunflower, a native of America was introduced as a commercial oilseed crop in India during the 1970's. Russian Federation, Ukraine and Argentina are the important sunflower producing countries in the world. In India, the crop has witnessed a considerable growth in terms of area and production since its introduction. It ranks fourth in area and ninth in production.

The relative photoperiod and thermo-insensitivity of the crop fits well in many multiple and intercropping systems in wide range of agroclimatic conditions both as catch crop as well as contingency crop. Karnataka, Andhra Pradesh, Maharashtra and Tamil Nadu are the principal sunflower growing states called as traditional areas accounting for more than 90% of crop being grown in *kharif* and *rabi* seasons. *Rabi* sunflower area is increasing in recent times. Since the nineties, spring season sunflower in the Indo-Gangetic region of Punjab, Haryana, Western Uttar Pradesh, Bihar and West Bengal is of importance for their high productivity, due to the association of positive effects of crop rotation, high soil fertility, irrigation and ideal growing conditions.

The crop has witnessed a considerable growth in terms of area and production since its introduction. From a meagre 3.09 lakh ha (1970–85), it has touched 18.00 lakh ha mark during (1994–03) which resulted in 475% increase in production with the productivity showing a declining trend. Among all the nine oilseed crops, its annual average compound growth rate in production was 14.9%, next to soybean (22.82%). In terms of area, the annual growth rate of this crop was next to sesame. In 2002–05, it's growth rate of productivity is –0.09%. During the period 1970–85, there was negative growth rate in productivity (–3.64%) which significantly improved during 1985–94 (+6.04%) but the area expansion was much faster during the pre-Technology Mission period. The export of sunflower seed extracts also showed a declining trend (123.04 thousand tonnes in 1994–95 to 2.8 thousand tonnes in 2005–06) (Fig 7.1).

Achievements in the Development of Hybrids

Leclercq (1969) discovered cytoplasmic male sterility in sunflower from the cytoplasm source of *Helianthus petiolaris*. Cytoplasmic male sterility and fertility restoration system has been extensively used to produce hybrids in sunflower. About 40 different cytoplasmic sources have been identified and characterized in sunflower but still *pet 1* cytoplasmic source is being used in commercial hybrid development programmes. The following are the procedures recommended for a successful maintenance of inbreds and commercial hybrid seed production.

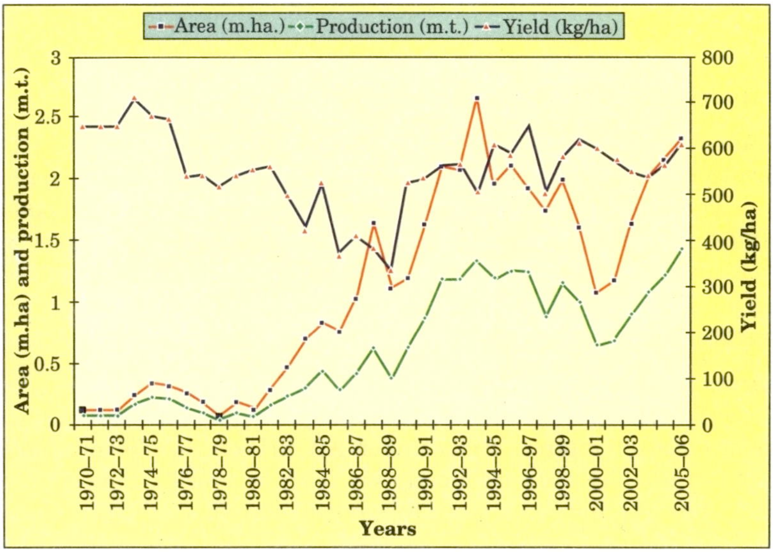

Trends in area, production and yield of safflower in India

Figure 7.1

Maintenance of Parental Lines

It is essential to maintain the purity and vigour of the parental line to realize the potential yields from the hybrids and to avoid rapid genetic deterioration of the hybrids. Initially B and R lines are grown in isolation and 1000 selected plants conforming to the standard in each group is selfed and based on productivity 200 plants are selected. Plant to row progenies of these plants is raised while a portion of the selfed seeds is retained as remnant seed. The lines not conforming to the standards are rejected. Identify 50 best progeny rows based on yield, oil and other agronomic characters and the ramnant seeds are bulked and grown in isolation. Ensure adequate cross pollination and bulk the seeds to form the nucleus seed of B or R line. Similarly A lines are maintained by pair crosses with B lines conforming the standards. The seeds of pair crosses and its respective B plants are grown and evaluated while retaining a portion of remnant seeds. Based on the presence of pollen shedders and other standards A and B plants are selected and their remnant seeds are bulked separately to form the A and B line nucleus seeds.

To take up breeder seed production of B and R lines, nucleus seeds are used and about 1000 M isolation distance is maintained to avoid contamination. Good cross pollination is ensured and breeder seeds are collected separately. Whereas in A line, a planting ratio of 3 :1 or in separate blocks in a ratio of 75:25 of A and B lines are followed. Manual transfer of B pollen to A line ensures good seed set. Removal of B plants immediately after pollination period is followed to avoid mechanical mixtures during harvest.

Commercial Hybrid Seed Production

Foundation seed of A and R lines is to be planted in the ratio of 3:1 or two separate blocks. Isolation from other sunflower crops should be at least 400 m. The pollination in sunflower is aided by insects and wind pollination is almost negligible. Hand pollination is essential to ensure good seed set thereby increased hybrid seed production. Removal of R plants immediately after pollination period ensures the purity of hybrid seeds by avoiding physical mixtures.

After the release of BSH-1, the first sunflower hybrid in India in 1980, several hybrids have been successfully developed and released for cultivation after intense trials under AICRIP on Sunflower (Table 7.2). The public bred hybrids are mostly single cross hybrids. However, private firms are marketing three way cross hybrids to the farmers.

Table 7.2. Sunflower hybrids released for cultivation by public institutes

Hybrid	Year of release	Yield (kg/ha)	Days of maturity	Oil (%)	Area of adaptability	Salient features	Pedigree
BSH-1	1980	900	85	41	All India	Resistant to rust & downy mildew	CMS 234A × RHA 274
APSH-11	1987	1200	90–95	41	AP		CMS 7-1A × RHA 271
LSH-3	1988	1200	95	39	Maharashtra	Resistant to downy mildew	CMS-207a × MRHA-1
KBSH-1	1992	1400	90	43	All India	High oil	CMS 234A × 6D-1
PSFH-67	1993	1300	95	41	Punjab		CMS 234A × P35R
PKVSH 27	1996	1300–1400	85–90	39	Vidharba region of Maharashtra	Moderately resistant to downy mildew	CMS2A × AK-1
DSH-1	1997	1300	85	40	North Karnataka	Resistant to downy mildew	DSF 15A x RHA 857
TCSH-1	2000	1500	85–90	39	Tamil Nadu	Tolerant to *Alternaria* & rust	CMS 234A × RHA 272

Table 7.2. *Contd.*

Hybrid	Year of release	Yield (kg/ha)	Days of maturity	Oil (%)	Area of adaptability	Salient features	Pedigree
KBSH-41	2002	1300–1500	90–92	39–41	Karnataka	Tolerant to moisture stress	CMS-234 A × RHA 95 C-1
KBSH-42	2002	1300–1500	90–92	38–41	Karnataka	Tolerant to moisture stress	CMS 851A × RHA 95 C-1
PSFH-118	2002	1400	85–88	40	Punjab	Resistant to stem and head rot. Tolerant to *Heliothis* and white files	CMS 10A × P61R
KBSH-44	2003	1400–1600	95–98	36–38	All India	Resistant to downy mildew	CMS 17A × RHA 95 C-1
LSFH-35 (Maruti)	2003	1400–1500	Medium	39–41	Maharashtra	Resistant to downy mildew	CMS-235 × RHA-1-1
NDSH-1 [NDSH-15]	2003	1400	88–90	40	Andhra Pradesh	Early maturing hybrid	CMS-234A × 859R
RSFH-1	2004	1300–1500	95–100	39–41	Karnataka (Zone 2 and 3)	...	...
HSFH-848	2005	1800–2400	90–95	41–42	Haryana	...	...
DRSH 1	2006	1300–1600	92–98	42–44	All India	High oil	ARM 243A × 6D-1

Safflower

Safflower crop is valued for its high quality edible oil. India has the highest acreage under safflower and stands first in area and second in production after Mexico in the world. Highest productivity of safflower is recorded in Mexico (2503 kg/ha), China (2333 kg/ha) and USA (1441 kg/ha). There is declining trend in area and production of safflower in the world mainly due to the situation in India.

In India, safflower is more remunerative than most of the *rabi* crops in drought prone areas as the crop has specific adaptation to *rabi* season and for residual moisture conditions in Vertisols. Parts of Maharashtra, Karnataka and Andhra Pradesh are the traditional safflower areas. The crop has high potential in many nontraditional areas like Gujarat, Madhya Pradesh and West Bengal. Despite the market demand, the area and production of safflower have been steadily declining since the late eighties mostly due to the availability of sunflower and palm oils at cheaper rates. The crop has high potential for export due to high nutritional and pharmaceutical properties of oil. The spiny nature of the crop makes it less amenable for newer areas for adoption.

The crop had recorded a highest area of 1 million ha during 1988–89 from an acreage of 0.15 million ha during 1965–66. Similarly, it recorded improvement in production and productivity during Technology Mission period. The current situation of safflower is lowest and showing declining trend. There was steep reduction in export of safflower seed. During 1965–85, safflower performed better in area, production and productivity, the corresponding annual growth rates being 3.10, 7.58 and 10.91%, respectively. But during the post-Technology Mission period, it lost area (–4.41% per annum) and registered a lower growth rate in production (0.01%). However, the growth rate in productivity (4.63%) was better than groundnut and rapeseed-mustard. During 1965–85, the crop area was 6.48 lakh ha which increased to 8.11 lakh ha in 1985–94. Similarly, it recorded improvement in production and productivity during the post-Technology Mission period over 1965–85. The country had earned Rs 475 lakhs through the export of safflower product (27.67 lakh MT) in 2005–06.

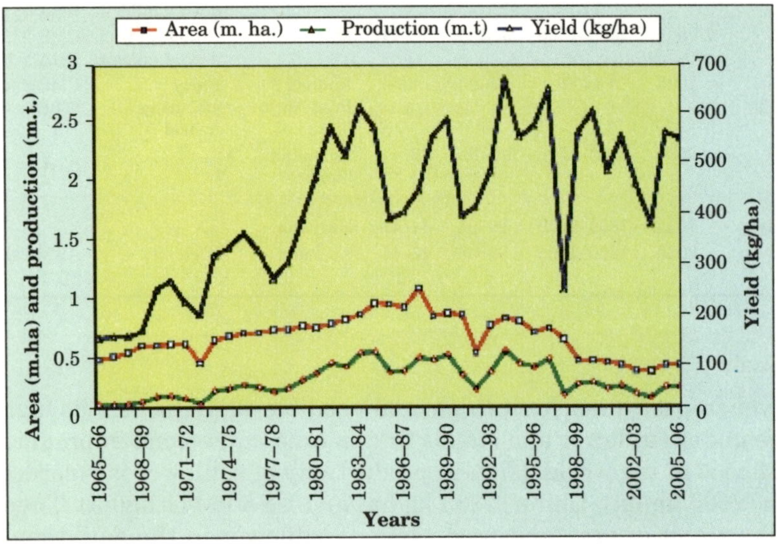

Trends in area, production and yield of safflower in India

Figure 7.2.

Achievements in the Development of Hybrids

The heterotic potential of safflower crop has amply been demonstrated. Both GMS and CMS systems are available in safflower. However, only GMS has been successfully exploited to so far. The recent developments in the search for stable cytoplasmic male sterility and fertility system in safflower have raised the hope to have a more economically viable and practically feasible seed production in safflower. The following are the procedures recommended for a successful maintenance of inbreds and commercial hybrid seed production.

Maintenance of parental lines: The genetic male sterility system being exploited for hybrid development in India is of both monogenic recessive and dominant nature. Nucleus seed production of genetic male sterile line determines the purity and quality of seed production. Pair-wise selection of male sterile and fertile plants with similar characteristics of original is carried out and crossed. At least 50 such pairs are made under caged condition. The harvested seeds from each plant are raised in progeny rows along with the original MS line. The uniform similarity for different traits with check and segregating 1:1 MS and MF are controlled pollinated with fertile sib counterparts. The male sterile plants thus selected are harvested and threshed separately and the seeds of progenies exhibiting similarity with the original MS plant are bulked to form the nucleus seed. Safflower is usually a self-pollinated crop, The pure lines are used as male parent for hybrids and maintained by following the standard procedure. The breeder seed of GM plants is produced by sowing the nucleus seeds. The GMS line will segregate as 1:1 MS and MF plants. The MS plants are identified at capitulum opening stage and marked. Optimum honeybee activity at flowering is desirable to obtain maximum seed yield. The MS plants are harvested first to avoid physical mixture.

Commercial hybrid seed production: An isolation distance of one kilometer is essential to avoid pollen contamination. Six rows of GMS lines and two rows of male parent are planted in the seed production plot. The MS plants are identified at flowering stage and the MF plants alone rouged out by observing the capitula 5 to 6 days prior to opening and the rouging is completed before the start of flowering. Optimum honeybee activity at flowering is desirable to obtain maximum seed yield. The MS plants are harvested first to avoid physical mixture with male parent.

India is the first country to release a hybrid in safflower and at present three public bred hybrids based on GMS system are popular (Table 7.3). The recent developments of CMS hybrids are likely to make a major impact on crop productivity and its competitiveness in the coming years.

Table 7.3. Safflower hybrids released for cultivation by public institutes

Hybrid	Year of release	Yield (kg/ha)	Days of maturity	Oil content (%)	Area of adaptability	Salient features	Pedigree
DSH-129	1998	1989	130	31	All India	Moderately tolerant to wilt, 25% high yield and 20% high oil yield	MS9 (0) × A1
NARI-NH-1	2002	1936	135–142	31	All India	Non-spiny, tolerant to *Alternaria* and aphids	MMS × C 2829-5-3a-6
NARI-H-15	2005	2200	129	29	All India	—	—

Castor

Castor is a major oilseed crop with unique property of ricinolic acid for industrial use mainly as lubricant and further through their innumerable derivatives for speciality industrial and pharmaceutical purposes. India is a major player in world castor production. Castor area in the world is declining since 2000 after attaining the highest area of 1.83 million ha. In India, the phenomenal increase in production of castor from 2.08 lakh tonnes to 5.06 lakh tonnes was possible due to significant increase in productivity (74%) with 40% of area increase. Among all the oilseed crops, the annual growth rate in productivity of castor was the highest during 1985–94 (12.11%) and in production it ranked third (15.42%), next to soybean and sunflower.

Gujarat, Andhra Pradesh and Rajasthan are the main castor producing states. The productivity is highest in Gujarat and Rajasthan, being cultivated under high input irrigated environment. The monopoly of India as sole castor producing country may be challenged with China and Brazil coming up in a big way in castor production. Under such situations castor prices may crash due to saturation of its industrial requirements. The efforts to develop ricin free castor in USA may give a fillip for castor production in that country in the near future.

The introduction of hybrids and the management technologies generated has resulted in increased area, production and productivity of castor under irrigated conditions of Gujarat and Rajasthan. In rainfed areas, high yielding and early maturing varieties gave high and stable yields over the years. The increased production helped in earning valuable foreign exchange. This crop was introduced into the non-traditional areas of Rajasthan and coastal Andhra Pradesh in different cropping systems. There is a need for development of wilt resistant hybrids and technology for higher water use efficiency (Fig 7.3).

Achievements in the Development of Hybrids

The first castor hybrid was developed and released in India. Unique manipulation of sex expression in the monoecious inflorescence has been the highlight of castor hybrids. The femaleness in castor is polygenically controlled and highly unstable. Being a cross pollinated crop wind and insects play a major role in aiding pollination. Hence, an isolation distance of 1500–2000 m is recommended for maintenance and seed production.

Maintenance of parental lines: The selfed seeds of individual plants of female line are grown in summer with a mean temperature above 32°C. This allows maximum expression of environment sensitive staminate flowers in pistillate raceme. Depending on uniformity, 10–15 plants per progeny are retained and others removed. The plants which are pistillate

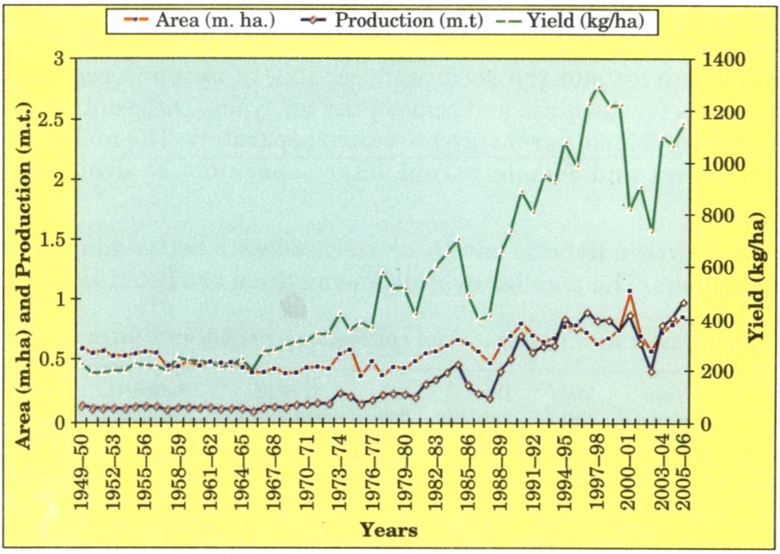

Trend in area, production and yield of castor in India

Figure 7.3.

up to fifth order only are selected. The selfed seeds are harvested plant wise and progeny wise and the seeds of all selected open pollinated racemes are harvested plant wise. The deviants based on 100 seed weight and oil content are rejected and the open pollinated seeds of selected selfed plants are bulked to form the source for breeder seed production while the seeds of selfed racemes are used for maintenance and nucleus seed production. Similarly in the male line, which is nothing but an open pollinated variety, about 50–100 selected and selfed progenies are grown for maintenance and initial multiplication of nucleus seed. Two to five hundred plants grown in progeny rows are evaluated and morphologically uniform plants at the primary spike stage are retained. Five to 10 plants in each selected progeny are labeled and selfed leaving primary spike open. Based on morphological data from open pollinated spikes progenies showing least intra-progeny variation is selected. About 50 seeds from each plant from the selected progenies are used as a source for further maintenance and nucleus seed production and the remaining portion of the selfed seed used as stock for breeder seed production.

Commercial Hybrid Seed Production

Hybrid seed production in castor is a highly skilled and specialized process because of the typical breeding system. The planting ratio of 3:1 female : male is followed. A minimum of 150–300 m isolation should be maintained. The best period for planting for hybrid seed production is August first fortnight to end in Gujarat and September second fortnight to October first fortnight in the southern states. Any change in these schedules will have undesirable effect on sex expression of female parent. Rouging is the most important activity and the seed producer has to examine regularly the female plants for deviants and remove the off types. Different sequential order spikes of female parent are harvested separately. The male parent is harvested first and female parent later separately to avoid physical mixtures.

So far fourteen hybrids have been released with better adaptability to various regions. The popular hybrids among them are listed in Table 7.4.

Table 7.4. Castor hybrids released for cultivation by public institutes

Hybrid	Year of release	Yield (kg/ha)	Days of maturity	Oil con-tent (%)	Area of adaptability	Salient features	Pedigree
GCH-3	1986	1700	200–210	48	Gujarat	For irrigated condition	VP-1 × JI 35
GCH-4	1988	2200	150–180	50	Gujarat	Dry & irrigated conditions	VP-1 × 48-1
GCH-5 (SHB-145)	1996	1740 (rf)/ 2820 (i)	120–240	50	All India	Resistant to wilt, jassids	Geetha × SH-72
DCH-32 (Deepti)	1998	1800 (rf)/ 2800 (i)	100	49	All India	Resistant to sucking pests & tolerant to wilt	LRES-17 × DCS-5
GCH-6 (JHB-665)	1999	1398 (rf)/ 2323 (i)	120–240	48	Gujarat, Rajasthan, Maharashtra	Resistant to root rot and white flies	JP 65 × JI 96
DCH 177 Deepak)	2000 2000 (I)	1500 (rf)/	90–180	49	Rainfed areas of Deccan plateau and irrigated areas of Maharashtra	Resistant to white fly, Resistant to wilt	DPC-9 × DCS-9
DCH-519	2004*	1736	105–110	49	All India	Resistant to *Fusarium* wilt, leaf hoppers	M574 × DCS-78 VP-1) × (163-1-11 × 150-4)
PCH-1	1999	2000 (rf)	180	48	Rainfed areas of AP	Resistant to leaf hoppers	VP-1 × PCS 124
TMVCH-1	1999	1403	90–170	52	Tamil Nadu	Moderately resistant to wilt and gray mold	LRES-17 × TMV-5
RHC-1	2002	2000(i) 1200 (rf)	180	48	Irrigated and rainfed areas of Rajasthan	Tolerant to wilt and resistant to jassids	Vp-1 × TMV5-1
GCH-7	2007	3500	180–210	48–49	Gujarat	Resistant to wilt and tole-rant to white fly, jassids and thrips	SKP 84 × SKII 215

Scope and Importance of Hybrids

Currently, sunflower, safflower and castor account for around 10.5 and 8.4% of the total oilseeds area and production in the country, respectively. Since the demand for oilseeds is sharply increasing there is urgent need to increase the domestic production. There is substantial potential to increase the production by increasing the area under sunflower, safflower and castor in the country. By 2025, the area under sunflower, safflower and castor can be increased to at least 4.5, 1.0 and 1.1 million hectares by extending the cultivation of these crops into non-traditional areas and seasons, replacement of less remunerative crops, intercropping with compatible crops, etc. Since this option is not practically possible in the present context, the other viable option is to increase the productivity of these crops by genetically synthesizing favourable gene combinations in the form of high yielding varieties/hybrids and overcoming the production constraints in various crop growing situations. Enhanced productivity coupled with area expansion of these crops must be able to contribute about 20% of total annual oilseeds production by 2025. Thirty one hybrids and 17 varieties in sunflower, 5 hybrids and 20 varieties in safflower, and 14 hybrids and 18 varieties in castor are released and notified for cultivation through the AICRP network. While hybrids dominate in sunflower and castor arena, varieties continue to meet the demand of safflower. Since the potential of open pollinated varieties in the case of sunflower and castor which are predominantly cross-pollinated and the pure lines of often cross pollinated safflower have reached a stagnation beyond which no further improvement is predicted the realistic option left with any breeder is to look into the heterotic potential of the new genetic combinations.

Among the oilseed crops, castor, sunflower and safflower have the possibility of exploiting heterosis resulting in quantum jump in productivity through breeding efforts. Several studies have brought out the feasibility of exploiting hybrid vigour in these crops resulting in the development of several hybrids which are superior in terms of seed and oil yields. It is interesting to note that different mechanisms have been exploited to produce hybrid seeds in larger scale and the systems were proven to be economically viable.

Future Strategies for Hybrid Development

Sunflower being a photoperiod insensitive and short duration crop has an unbound future to expand both in horizontal and vertical directions and dominate edible oilseed scenario of the country. The interest in safflower is rejuvenating due to the pharmaceutical uses of the petals, which could also act as a source of natural dye. The recent success in development of CMS hybrids has ushered in an opportunity to achieve quantum jump in safflower production. Castor crop continues to rule as the unparalled source of industrial oil with multiple uses. The declining *per capita* availability of land for cultivation, unstable productivity under rainfed conditions, competing crops and priorities will pose a problem for expansion of area under these crops. Hence, the barriers for achieving higher productivity like limited genetic variability among the parents/inbreds, susceptibility to biotic and abiotic stresses, etc., need to be addressed by the use of cutting edge technologies. Some of researchable issues which need to be addressed to achieve quantum jump in productivity of these hybrid crops are as follows:

- Development of sunflower hybrids with resistance to *Alternaria,* and necrosis (SND) with high oil content (44–46%).
- Development of suitable short duration sunflower hybrids for spring season cultivation in northern India.
- In safflower, identification of efficient cytoplasmic genetic male sterility system to exploit hybrid vigour and standardizing appropriate hybrid seed production technology.
- Development of hybrids with high oil content (33–35%) and resistance to wilt, *Alternaria* and tolerance to aphids in safflower.
- Development of short and medium duration hybrids with resistance to *Botrytis,* wilt complex and capsule borer in castor.
- Maintenance and upgradation of seed yield, oil content and disease resistance of parental lines has to be a continuous effort in these crops.
- Generation of pre-bred germplasm through introgression of desirable traits like resistance to *Alternaria* from wild sunflower and safflower to widen the genetic base of the hybrids/inbreds and development of recombinant inbred lines (RILs) and near isogenic lines (NILs) with contrasting characters.
- Concerted efforts should be made to produce adequate quantities of certified seeds of the promising hybrids to cater the need of the farmers.

Strategies for Commercialization

Sunflower, castor and safflower are important oilseed crops where the phenomenon of heterosis has been exploited extensively to produce hybrids on commercial scale. Several hybrids have been bred and released for cultivation both by public research organizations and private sector companies utilizing the cost effective cytoplasmic male sterility-fertility restoration system in case of sunflower, genetic male sterility in safflower and unique pistillate lines in castor. However, private hybrids are also popular among the farmers since the timely availability of seeds is ensured by the seed firms. Though many of the public bred hybrids listed earlier are more promising in terms of seed and oil yield and were proved to be superior over private owned hybrids in the All India Trials, the availability of adequate seed in terms of quantity and quality is a major limiting factor for their spread in major traditional growing regions.

Every care has been taken to popularize the hybrids released based on their performance in trials through demonstration at farmers fields. The results obtained so far from the project on "Frontline Demonstrations in Oilseed Crops" sponsored by the Dept. of Agriculture and Cooperation, GOI under the Oilseeds Production Programme (OPP) have conclusively proved the potential of improved hybrids developed by DOR and its centres over the prevailing varieties of farmers' choice. However, the inadequate availability of certified seeds in time and quantity is a bottle neck. Hence, they were forced to look into private hybrids, seeds of which are readily available. As seed production is dominated by private sector the cost of seed is also very high. Besides, the threat of spurious seed being sold by fly-by-night companies increases the risk. Aggressive quality seed production and quality control by governmental agencies makes hybrid seed available at cheaper cost and assures quality. To facilitate effective quality seed production of hybrids the following strategies may be considered.

- Since sunflower and castor are highly cross-pollinated, seed production of both parental lines as well as hybrids requires adequate isolation to produce genetically pure seed by avoiding contaminations. In such a case, production of large quantity of quality hybrid seeds in research farms is not feasible.
- One strategy to overcome the problem of isolation is to grow the seed crop in nontraditional season. For instance, to produce hybrid castor seed of DCH 519, a recently released hybrid from DOR, seed production was planned and successfully completed this year in areas like Challapur of Kosgi and Chennaram and Masaipalli of Daulathabad mandal of Mahabubnagar district where castor is not grown this season. In some cases even switching over to the areas where the crop is not grown normally may also help. Similarly off-season niches can also be explored to produce hybrid seeds. For example, the off season facilities at Wellington for winter crops may be utilized for safflower hybrid seeds production.
- DOR has also evolved a strategy to engage farmers in seed production of hybrid castor so as to produce quality seeds in larger quantities and in turn farmers get remunerative price for their products. This has in a way helped to overcome the problem of isolation and land requirement while resulting in easy dissemination of the hybrids developed by the Institute. When a large number of farmers in a particular location are involved in such activity of seed production of

a single hybrid, the problem of isolation and contamination is avoided. It is only successful when the stray castor plants along the road sides are removed carefully and all the farmers in that area cooperate to grow only the suggested parental lines.

- The seed village concept has been well planned and executed by DOR in case of hybrid castor seed production. However, the main responsibility of maintaining the genetic purity of the parental lines (nucleus seed) should lie with the Breeder or the Institute who developed the hybrid. Further seed increase in the form of breeder seed may also be done at the institute and the certified seed production may be taken up at the farmers' field.
- The farmers should be trained in all the aspects of hybrid seed production, *viz.*, identification of off-types, rouging, other special aspects to get quality seed, crop management, protection against pests and diseases, etc.
- In any of these innovative programmes on seed production the involvement of the seed certification agencies must be ensured.
- DOR was successful in demonstrating the seed village concept for the past several years. Inadequate marketing support is a major problem in these attempts. By this cooperative farming method the farmers themselves sell the hybrid seeds to other farmers. However, if the quantity produced is large it is always better if a tie up is planned with some of the seed agencies and NGOs. For instance, at the seed village planned at Lingampalli of Manchal mandal and at Chellapur in Kosgi mandal, NGO like WASSAN and self-help groups (SHGs) have been involved in marketing of the hybrid seed from seed growers to the farmers by mutual sharing of benefits.
- The economics of seed production of castor hybrid DCH 519 from Lingampalli village of Manchal mandal revealed that the net profit occurred was Rs 9700/acre and the B:C ratio was 3.9 though the crop suffered severely from biotic stresses. Thus, the economic security of the farmer is taken care of. This works well in small scale but the success when it involves a larger area and quantity needs to be verified.
- India has adopted a legislation through the Protection of Plant Varieties and Farmers' Right (PPV&FR) act in 2001. The act has been enacted in order to provide for the establishment of an effective system for protection of plant varieties, the rights of farmers and plant breeders and to encourage the development of new varieties of plants. The candidate variety should be distinct from existing varieties, uniform in its essential characteristics and stable in essential characteristics after repeated propagation. Hence, the first step before privatizing seed production of public bred hybrid is to register the

parental material. The genetic identity of lines registered should also be documented and supported by molecular data.

- To popularize the public-bred hybrids among farmers by providing adequate quality and quantity of seeds, it is also advisable to go for licensing of the parental inbred lines to private seed producers so as to ensure timely seed availability to the growers by following stipulated guidelines. It should be a condition that the seed production plots raised from parental seeds from public sector should be evaluated by the breeder concerned at critical stages to avoid any problems related to genetic purity. It is also essential to insist in the MOU that the name of the public hybrid is retained by the private firms while marketing.
- The PPV & FR Authority has opened up the registration/licensing only to 13 crops as on today in which sunflower, safflower and castor are not included. However, it is likely that registration/licensing will be opened up for sunflower and castor also from June 2008 onwards. Hence, it is suggested to wait till that time before entering into any agreement with private firms regarding licensing of inbreds/parental lines to produce public bred hybrids of sunflower, safflower and castor on commercial scales.
- Popularization of hybrid technology results in higher productivity per unit area and increased acreage under hybrids leading to enhanced production of oil. The import of edible oils will be minimized saving valuable foreign exchange. Economically viable hybrid technology is not only helping in food security. It also ensures the availability of nutritionally rich edible oil from safflower and sunflower which are preferred by health conscious people.

8

Breeding Hybrids for Enhancing Productivity in Pigeonpea

K.B. SAXENA*

ABSTRACT

The development and promotion of pure line varieties has helped in stabilizing the productivity and in increasing production of pigeonpea at the national level without any sign of improvement in productivity. The advent of CMS based hybrid technology may achieve a quantum jump in the productivity of pigeonpea. More resources are required to develop the parental materials suitable for diverse environments with insecticidal genes combined with high grain quality to show the impact of hybrid technology in farmers' fields. To enhance the pace of research and development of hybrid pigeonpea both public institutes and private sector seed companies are sensitized to the technology with sharing of knowledge, breeding materials, and experience under the concept of "consortia". Behaviour of plant, male sterility system, development of hybrids, diversification of parents, disease resistance, seed production technology, economics of hybrid, and new frontiers and strategies are described.

Key words: Heterosis, hybrid pigeonpea, male sterility system, out-crossing, seed production

The nature and action of genes determining yield in cereals and pulses are more or less comparable. But the quantum of yield advances made through breeding in grain legumes is far less than in cereals, and in particular, much less than that made through the commercial exploitation of hybrid vigour in cereals. Although floral biology of a few legumes permits limited out-crossing but it was never used in the past to develop commercial hybrids. ICRISAT took initiative in this direction and developed an efficient hybrid pigeonpea breeding technology. This technology is based on insect-aided natural out-crossing and a cytoplasmic-nuclear male-sterility system that was developed by integrating the cytoplasm of *Cajanus cajanifolius*

* Principal Scientist, International Crops Research Institute for the Semi-Arid Tropics, Patancheru 502 324, A.P.

(a wild relative of pigeonpea) and the nucleus of a cultivated variety. Some of the recently developed experimental hybrids have recorded 50–100% yield advantages over the popular varieties. This paper discusses the achievements and prospects of commercializing pigeonpea hybrids.

INTRODUCTION

Pigeonpea [*Cajanus cajan* (L.) Millsp.] is a versatile plant species which can grow successfully in a range of soil types, temperatures, and photoperiods. Its deep and strong root system enables it to overcome intermittent droughts and other stresses. Pigeonpea is globally cultivated on 5.25 million hectares with an average production of 3.2 million tonnes. It is grown mainly in semi-arid regions of Asia, Africa, Latin America, and Caribbeans. In India it is predominantly grown in the states of Maharashtra, Uttar Pradesh, Karnataka, Andhra Pradesh, Bihar, and Gujarat. The adoption of new varieties in India has helped in increasing pigeonpea area from 2.3 million hectares in 1950 to 3.58 million hectares in 2005. However, no increase has been witnessed in the productivity of the crop that has remained unchanged at about 700 kg ha^{-1} and the country is still short in supply of grains by a margin of over 500,000 tonnes annually.

The quantum jump in yield potential observed in some cereal and vegetable crops in the past was primarily due to exploitation of hybrid vigour. In pulses, however, this breeding approach could not be used due to their highly self-pollinating nature and inability to produce large quantities of hybrid seed economically. Pigeonpea is a unique crop where a reasonable extent of natural out-crossing occurs and ICRISAT scientists believe that this trait could be used in breeding commercial hybrids. For a successful commercial hybrid breeding, it is imperative that the hybrids have significant yield advantages over the existing varieties, and their seed production is cost-effective. To achieve this, the most crucial factors that are needed are (i) a stable CMS system, (ii) diverse fertility restorers, and (iii) a mass pollen transfer mechanism. In pigeonpea, all these components are now available and this has made the commercial hybrids a reality.

Natural Out-Crossing

The efficiency of any hybrid seed production programme depends on the effectiveness of controlled natural mass pollen transfer mechanism from the male to female parent. This ensures economical production of quality hybrid seed. Williams (1977) reported that *Apis mellifera* and *Megachile* species of insects play an important role in natural out–crossing in pigeonpea. In the process of collecting nectar from flowers large quantities of pollen get attached to their bodies which effect cross-pollination. On an average, the extent of natural out-crossing is about 25% but it may vary in either direction depending on the intensity of insect population. A review conducted by Saxena *et al.* (1990) revealed that in most places where pigeonpea is grown the natural out-crossing occurs and the level of out-crossing is sufficient to undertake an effective hybrid breeding programme.

Cytoplasmic-Nuclear Male-Sterility (CMS) Systems

The cytoplasmic nuclear male-sterility (CMS) system is the most widely accepted model of producing commercial hybrids in field crops. The expression of CMS, in part, is controlled by the factors carried only through the female parent, which is never lost or diluted in the succeeding generations. Nuclear genes generally influence the expression of this trait. The male-sterility effects are conditioned by interactions among various nuclear and cytoplasmic factors. The cytoplasmic factor is referred to as 'N' for the normal fertile cytoplasm and 'S' for the sterile cytoplasm. The CMS lines are maintained by these two cytoplasmic factors. The CMS (A-line) must have 'S' cytoplasm and nuclear gene s for *'msms'*, and the maintainer B-line must have a normal cytoplasm with *'msms'* nuclear gene. The F_1 between A and B-lines, therefore, is always male-sterile because the 'N' cytoplasm responsible for fertility in the B-line is not transferred to the F_1. For producing fertile hybrid seed, the A-line with 'S' cytoplasm is crossed with a fertile line with fertility restorer nuclear genes, commonly known as *'Fr'* genes. To sum up, the three-line system is geared for multiplying A-line with the help of B-line and for producing the hybrid seed, the A-line is crossed with R-line.

The development of pure breeding CMS system in pigeonpea would facilitate the commercial seed production. Since the search for a functional male-sterility in the germplasm succeeded in identifying only genetic male-sterility, the efforts were made to develop CMS through planned plant breeding. The first such unsuccessful attempt was made by crossing *C. scarabaeoides,* a wild relative of pigeonpea, with a cultivated type (Reddy and Faris, 1981). Ariyanayagam *et al.* (1995) crossed *C. sericeus* with a cultivated type and in the backcross populations male-sterility was observed. The reversion of some male-sterile plants to fertility always complicated the selection and stabilization of this trait. Intensive selection in additional backcross generations, however, resulted in the identification of a promising CMS line (Saxena *et al.*,1996). This was designated as A_1 cytoplasm. The scientists at Gujarat Agricultural University and ICRISAT simultaneously developed a CMS system with *C. scarabaeoides* as cytoplasm donor (Tikka *et al.*, 1997; Saxena and Kumar, 2003) and it was designated as A_2 cytoplasm. At Akola, another CMS system was developed (Wanjari *et al.,* 2001) from one more interspecific crosses involving *C. volubilis* (A_3 cytoplasm). The recently developed CMS system, designated as A_4 cytoplasm (Saxena et.al., 2005), has the cytoplasm of *C. cajanifolius*, another close wild relative of pigeonpea (Table 8.1). The male-sterile plants in this material show no morphological deformity and produce plenty of pollen grains in hybrid combinations with fertility restorers. This male-sterile source is stable and capable of producing high yielding hybrids (Saxena *et al.*, 2006). Therefore, it has a good potential in commercial hybrid breeding.

Table 8.1. Segregation of male-sterility over generations in A_4 CMS system

Generation	Total plants	Fertile
BC_7F_1	17186	0
BC_6F_1	1133	0
BC_5F_1	67	0
BC_4F_1	7	0
BC_3F_1	165	0
BC_2F_1	5	1

High Yielding Hybrids

In 2005 season, a number of short-duration hybrids involving A_4 cytoplasm were evaluated at five locations (Table 8.2). Of these, ICPH 2438 ranked first with mean yield of 2893 kg ha^{-1} and mean superiority of 106% over the control ICPL 88039 (1403 kg ha^{-1}). The other promising hybrids were ICPH 2363 (2447 kg ha^{-1}, 74% superiority), ICPH 2429 (2032 kg ha^{-1}, 45% superiority), and ICPH 2364 (1940 kg ha^{-1}, 38% superiority). In the medium-duration group also several hybrids were evaluated at four locations during 2005 (Table 8.3). Among these, hybrid ICPH 2744 (2790 kg ha^{-1}) at Patancheru, ICPH 2788 (4166 kg ha^{-1}) at Jalna, ICPH 3331 (3476 kg ha^{-1}) at Manoharabad and (3185 kg ha^{-1}) at Medchal were found promising. Overall, the promising hybrid combinations were ICPH 2788 (3263 kg ha^{-1}), ICPH 2740 (3066 kg ha^{-1}), and ICPH 3331 (2917 kg ha^{-1}). These hybrids exhibited 47–64% superiority (Table 8.3) over the control Asha (1989 kg ha^{-1}).

Table 8.2. Yield (kg ha^{-1}) of top five short-duration pigeonpea experimental hybrids in multilocation trials, 2005

Hybrid	Patancheru (ICRISAT)	Jalna (MAHYCO)	Mano-harabad (Zuari Seeds)	Medchal (JK Seeds)	Nagpur (Ankur Seeds)	Mean	Superiority over control (%)
ICPH 2438	3414	3792	1518	2236	3504	2893	106
ICPH 2363	2736	3958	1012	2060	2471	2447	74
ICPH 2429	2177	2625	1577	2024	1758	2032	45
ICPH 2364	2375	2300	1101	1696	2229	1940	38
ICPH 2460	2454	2708	1429	1238	1479	1862	33
ICPL 88039 (C)	1306	1750	1042	736	2179	1403	—
SEm (±)	472.2	NR	149.3	288.5	19 7.6		
CV (%)	28.2	9.5	16.0	24.4	12.1		

NR- not reported

Table 8.3. Yield (kg ha^{-1}) of top five medium-duration pigeonpea experimental hybrids in multilocation trials, 2005

Hybrid	Patancheru (ICRISAT)	Jalna (MAHYCO)	Mano-harabad (Zuari Seeds)	Medchal (JK Seeds)	Nagpur (Ankur Seeds)	Mean	Superiority over control (%)
ICPH 2438	3414	3792	1518	2236	3504	2893	106
ICPH 2728	2525	4166	3238	3124		3263	64
ICPH 2740	2761	4041	2524	2936		3066	54
ICPH 3331	2216	3791	3476	3185		2917	47
ICPH 2671	2671	3416	2571	2996		2913	46
ICPH 2751	2171	3125	2381	3493		2793	40
Asha (C)	1927	3124	1333	1570		1989	–
SEm (±)	207.7	NR	540.7	331.0			
CV (%)	13.7	7.3	32.2	21.1			

NR- not reported

SEED PRODUCTION TECHNOLOGY OF HYBRIDS AND THEIR PARENTS

Isolation specifications: Since the extent of natural out-crossing in pigeonpea is determined by the population of the pollinating insects, its extent may vary from one place to another. Therefore, it is difficult to specify isolation distances for different seed classes that would be effective at every location. We, however, need to develop some safe and standard guidelines to maintain the purity standards of the hybrids. Overall we feel that an isolation distance of 500 m will be sufficient for quality hybrid seed production in pigeonpea.

Seed production of A-lines: The nucleus seed of parental lines of hybrids should be produced with highest standards of genetic purity. For multiplying nucleus seed both A- and B- lines should be grown, preferably inside an insect proof cage. Each and every plant of A- and B-lines should be examined for various parameters and off-types be rogued and crossing should be done to produce the desired number of seeds. For the production of breeder seed of A-line, a field with appropriate isolation distance is selected and A- and B-lines are grown with recommended agronomic package. At ICRISAT, a ratio of 4 rows of A-line: 1 row of B-line has been found effective in producing seed of a given A-line. In the seed multiplication of ICPA 2039 at ICRISAT 27 kg of crossed seed was harvested from 225 m^2 block with an estimated yield of 1111 kg ha^{-1}. In another isolation of the same male-sterile line, 200 kg seed @ 877 kg ha^{-1} was harvested. The seed thus harvested be dried and packed after treating with Malathion powder @ 1 g kg^{-1} seed. In the short-duration materials, the mature pods on male-sterile and fertile plants be harvested by pod picking or by cutting the top pod bearing portion of the plants. The perennial nature of species will force the plants to re-generate and produce a second flush of flowers and pods.

Seed production of B- and R-lines: The maintainer (B) and restorer (R) lines are male-fertile in nature and their genetically pure seed can be produced without much difficulty. For nucleus seed production of B- and R-lines, about 100 plants are harvested from the central portion of the breeder seed production plot and their progenies be grown in the subsequent generation. After assessing their purity aspects, the selected progenies are bulked to serve as nucleus seed. For breeder seed production the nucleus seed should be multiplied in separate isolation. It is also important to multiply seed in large quantities so that foundation seed production is feasible. Always sufficient care should be taken to rogue out the off-types as and when such plants are identified in the seed production plot.

Seed production of hybrids: In a three-line hybrid seed production system, the hybrid seed produced by crossing A-line with R-line, is commonly called certified seed and it is grown on a larger scale. For the production of certified seed of hybrids the A-line and its pollen parent (R-line) are grown in 4 : 1 ratio in an isolated block. Some additional rows of pollen parent can also be sown on each side of the plot. This will enhance pollen availability for cross pollination. At flowering the pollinating insects will visit the male and female flowers in a random fashion and in the process collect pollen from fertile plants and carry out hybridization on the male-sterile plants. Using this technology in 2005 rainy season, from a small (120 m^2) isolation a total of 15 kg seed @ 1250 kg ha^{-1} of hybrid ICPH 2671 was produced. For optimum seed yields the male : female row ratio (1:4) need not be universally recommended for all the environments and it should be modified depending on the insect activity and time of sowing.

Cost of hybrids seed production: Seed cost plays an important role in adoption of hybrids. Both, the technology itself and crop management practices, are critical in determining the production costs. So far no study has been conducted to determine the production cost of CMS-based hybrids and their parents. A detailed cost of seed production study undertaken by Tamil Nadu Agricultural University, Coimbatore using the GMS-based hybrid system it was found that 813 kg ha^{-1} of hybrid seed was produced at the cost of Rs 6.25 kg^{-1} (Murugrajendran *et al.*,1990). In the CMS-based hybrid technology the rouging operation is eliminated and hence the production cost of both female parent and hybrid will certainly decrease by a significant margin. A detailed study on this aspect is needed.

Diversification of Hybrid Parents

ICRISAT has a strong hybrid pigeonpea breeding programme with increased emphasis on disease and insect resistance, stability and yield. At present we have 31 CMS lines with a large variation for important agronomic traits (Table 8.4). Seven of these A-lines also have high level of resistance to wilt and sterility mosaic diseases. We have also developed

91 fertility restorers in different maturity groups (Table 8.5). Of these, 37 restorers are resistant to wilt and sterility mosaic diseases. These genetic materials can be used to develop hybrids suitable for diverse environments.

Table 8.4. Variation for important agronomic traits among A-lines

Characters	Range
Days to flower	55 - 135
Days to maturity	100 - 198
Plant height (cm)	74 - 245
100-seed weight (g)	8.6 - 13.8
Seeds pod^{-1}	3.2 - 5.6

Table 8.5. Fertility restorers of A_4 cytoplasm available in different maturity groups at ICRISAT

Maturity group	Lines	Days to flower	Seed pod^{-1}	100-seed weight (g)
Extra-short	15	< 70	3.0–4.1	6.2–12.7
Short	17	71 - 80	3.0–4.1	6.7–12.1
Medium	45	81 - 130	3.4–4.5	7.5–14.7
Long	14	> 130	3.3–4.9	7.7–14.0

New Frontiers and Strategies

The development and promotion of pure line varieties has helped in stabilizing the productivity and in increasing production at the national level. But unfortunately, the productivity has shown no sign of improvement. The advent of hybrid technology has given us hope that the stagnated yield could be increased. To achieve a quantum jump in the productivity level of pigeonpea a good beginning has been made by ushering in hybrid breeding. The development of a stable CMS system in pigeonpea is a real boon to the breeders. The present achievements have established the hybrid technology but a lot needs to be done in the next decade to show the impact of this technology in farmers' fields and for this more resources will be required to develop the parental materials suitable for diverse environments. The next most important target will be to develop hybrids with insecticidal genes combined with high grain quality.

To enhance the pace of research and development of hybrid pigeonpea, ICRISAT is actively involved in technology transfer to national research system, including both public institutes and private sector seed companies. Now, our major responsibility is to take this product to our clients the farmers. We expect that both small- and large-holder farmers will show interest in the hybrids. However, since pigeonpea is predominantly cultivated by small, resource-poor farmers, we need to keep the seed costs within the reach of our clients. In this partnership the sharing of knowledge, breeding materials, and experience are our major strength.

The partnership between ICRISAT and private sector seed companies is strong and it has evolved over time. In the early years, ICRISAT played a nurturing role to the fledgling industry through informal networks. As private seed industry grew, it started to develop a significant research capability of its own and now they are our research partners–both as a source of funds and complementary expertise, especially in the area of development (Gowda *et al.*, 2004). Considering that each partner will contribute small amount of funds, ICRISAT initiated the concept of "consortia", under which a consortium member provides a small grant to augment research efforts. So far 12 private and one public sector seed companies (Table 8.6) have joined the Pigeonpea Hybrid Parents Research Consortium. Since last three years this consortium model is operating smoothly, and over 3000 seed samples of breeding materials and hybrid parents have been made available to the partners (Table 8.7).

Table 8.6. List of Pigeonpea Hybrid Parents Research Consortium members

Sr.	Name of seed company
1.	Ankur Seeds Pvt.Ltd., Nagpur, Maharashtra
2.	Biogene Agri-tech., Ahmedabad, Gujarat
3.	Bioseed Research India Pvt.Ltd., Hyderabad, Andhra Pradesh
4.	JK Agri Genetics Ltd., Hyderabad, Andhra Pradesh
5.	Krishidhan Seeds Ltd., Jalna, Maharashtra
6.	Maharashtra Hybrid Seeds Co. Ltd., Jalna, Maharashtra
7.	Maharashtra State Seed Corporation Limited, Akola
8.	Nath Seeds Pvt.Ltd., Hyderabad, Andhra Pradesh
9.	Nimbkar Seeds Pvt.Ltd., Phaltan, Maharashtra
10.	Nuziveedu Seeds Ltd., Secunderabad, Andhra Pradesh
11.	Pradham-Biotech Pvt Ltd., Hyderabad, Andhra Pradesh
12.	SM Sehgal Foundation, C/o ICRISAT, Hyderabad, Andhra Pradesh
13.	Zuari Seeds Limited, Bangalore

Table 8.7. Summary of pigeonpea hybrid breeding materials supplied by ICRISAT

Target organization	2006	2005	2004	Total
Private seed companies	1881	788	60	2729
Public seed companies	32	–	–	32
ICAR institutes	47	40	3	90
Agricultural universities	154	66	45	265

REFERENCES

Ariyanayagam R P, Rao A. N. and Zaveri P. P. 1995. Cytoplasmic-genic male-sterility in interspecific matings of *Cajanus*. *Crop Science*, **35**: 981–5.

Gowda, C.L.L., Reddy, B.V.S., Rai, K.N., and Saxena, K.B. 2004. ICRISAT collaboration with seed industry in Asia. Paper presented at Asian Seed Congress 2004, Hotel Lotle Jamsil, Seoul, Korea, September 13–17, 2004.

Murugrajendran, C., Rangaswami, S., Saxena, K.B. and Rangaswamy, P. 1990. Cost of hybrid seed production in pigeonpea. *International Pigeonpea Newsletter*, **11**: 12–13.

Reddy L.J. and Faris D.G. 1981. A cytoplasmic-genetic male-sterile line in pigeonpea. *International Pigeonpea Newsletter,* **1**: 16–17.

Saxena, K.B., Singh, L. and Gupta, M.D. 1990. Variation for natural out-crossing in pigeonpea. *Euphytica,* **39**: 143–8.

Saxena, K.B., Laxman Singh, Kumar, R.V. and Rao, A.N. 1996. Development of cytoplasmic-genic male-sterility (CMS) system in pigeonpea at ICRISAT Asia Center. Proc. Working Group on cytoplasmic-genic male-sterility (CMS) in Pigeonpea. 9–10 May 1996, ICRISAT Center: pp. 1–8. Patancheru, A.P. 502 324, India.

Saxena, K.B., and Kumar, R.V. 2003. Development of a cytoplasmic-nuclear male-sterility system in pigeonpea using *C. scarabaeoides* (L.) Thours. *Indian Journal of Genetics and Plant Breeding,* **63**: 225–9.

Saxena, K.B., Kumar, R. V., Srivastava, N. and Shiying, B. 2005. A cytoplasmic-nuclear male-sterility system derived from a cross between *Cajanus cajanifolius* and *C. cajan. Euphytica,* **145**: 291–6.

Saxena, K.B., Madhavi Latha, K., Kumar, R.V. and Dalvi, V.A. 2006. Commercial pigeonpea hybrids are just a few steps away. *Indian Journal of Pulses Research,* **19**(1): 7–16.

Tikka, S.B.S., Parmar, L. D. and Chauhan, R. M. 1997. First record of cytoplasmic-genic male-sterility system in pigeonpea (*Cajanus; cajan* (L.) Millsp.) through wide hybridization. *Guj. Agri. Uni. Res. J.,* **22**: 160–2.

Wanjari K.B., Patil, A.N., Manapure, P., Manjayya, J.G. and Manish, P. 2001. Cytoplasmic male-sterility in pigeonpea with cytoplasm from *Cajanus volubilis*. *Ann. Pl. Phy.,* **13**: 170–4.

Williams I. H. 1977. Behaviour of insects foraging on pigeonpea (*Cajanus cajan* (L.) Millsp.) in India. *Tropical Agriculture,* **54**: 353–63.

9 Strategies for Commercialization of Public - Bred Hybrids of Wheat in India

S.K. SINGH*, RAVISH CHATRATH* AND B. MISHRA*

ABSTRACT

Wheat hybrids can yield up to 30% more than their parents, but hybrids with heterosis at these levels usually are the product of crosses between different classes of wheat, such as a cross of hard red winter wheat with soft red winter wheat. Commercially useful wheat hybrids must be made within a class to maintain milling and baking quality. Crosses within a quality class typically have less heterosis because of relationship among members of a relatively closed gene pool. Cytoplasmic male sterility may allow production of hybrid in a perfect system of self-pollination. In India, hybrid development programme is addressed through CMS system in network mode. Pre-requisite for development of hybrid viz., diversification of male sterility system in indigenous background, identification of fertility restoration lines and their diversification and identification of heterotic groups, search and exploitation of open pollinating traits and standardization of the hybrid seed production technology for the development of a successful hybrid wheat technology are described.

Key words: Breeding strategy, CHA, commercial feasibility, hybrid wheat, male sterility system

GLOBAL WHEAT PRODUCTION SCENARIO

At global level, India ranks second largest wheat-producing nation and contributes approximately 11.7% to the world wheat production from about 12% of global area (USDA, 2007). China is the largest producer of wheat with 17.6% global wheat production from about 11% area under wheat. The other major wheat producing countries are the United States of America, the Russian Federation and Australia, and these five countries together contribute more than half of the global wheat production.

* Directorate of Wheat Research, Karnal 132 001.

Table 9.1. Major wheat growing countries and their production (million tonnes)

S. No.	Country	2003/04	2004/05	2005/06	2006/07	2007/08*
1.	China	86.49	91.95	97.45	104.47	106.00
2.	India	65.10	72.15	68.64	69.35	74.89
3.	USA	63.81	58.74	57.28	49.32	56.25
4.	Russia	34.10	45.40	47.70	44.90	48.00
5.	Australia	26.13	21.91	25.37	9.90	13.00
6.	Canada	23.05	24.80	25.75	25.27	20.60
7.	Turkey	16.80	18.50	18.50	17.50	15.50
8.	Argentina	14.50	16.00	14.50	15.20	15.50
9.	Iran	13.50	14.50	14.50	14.80	15.00
10.	Kazakhastan	11.00	9.95	11.00	13.50	16.00
	World	**554.42**	**626.83**	**621.66**	**593.66**	**602.31**

* *Estimations in December 2007*

The area under wheat throughout the world as well as in India has nearly stabilized around 215 m ha and 26 m ha, respectively. The enhanced genetic potential of new genotypes by introducing the semi-dwarf, photoperiod insensitive and fertilizer responsive varieties and increased area under cultivation during the Green Revolution era, resulting in increased productivity as well as production in India at the rate of nearly 1% per annum. After Green Revolution the *every* genotypes developed with 1B-1R translocation kept on increasing wheat yields in the country but the average production level has been hovering around 70 m tonnes during the last decade with the average productivity of 2.7 t/ha. As per the present population growth rate, population of India will be around 1.3 billion by 2025 AD. Assuming 20% more per capita requirement of food grains, due to better standard of living and increase in the demand of processing industries, wheat production must increase to around 109 mt by the year 2025 (DWR, 2007).

Scope and Importance of Hybrids

The wheat growing areas in the country have been divided into six mega zones based on the agro-ecological conditions. These are northern hill zone (NHZ), northwestern plains zone (NWPZ), northeastern plains zone (NEPZ), central zone (CZ), peninsular zone (PZ) and southern hills zone (SHZ). Among these, the northwestern plain zone (NWPZ) was the seat of green revolution and contributed maximum to the wheat basket of the country. Presently, it is being felt that this major wheat producing zone of our country comprising Punjab, Haryana, western UP and parts of Rajasthan has reached a sort of saturation level. Against the projected demand of 109 m tonnes by 2025 with productivity growth rate of 2.1% per year, India can harvest over 95 m t of wheat merely by bridging the present level of yield gap between the realized and potential yield as demonstrated through frontline trials especially in the northeastern plains zone comprising eastern UP, Bihar, Jharkhand and West Bengal and in central zone covering areas of MP, Chattissgarh, Gujarat and parts of Rajasthan and UP. Thus, there is need to explore new innovative approaches in order to break the yield barriers and make wheat cultivation more remunerative. In this context, exploiting hybrid vigour at commercial level through development of hybrid wheat is considered promising. Work on hybrid wheat started world over in 1962 and utilization of heterosis through hybrid breeding has given very high dividends in many field crops including horticulture.

The floral biology of wheat ensures that it is a 100 per cent self pollinated crop with occasional outcrossing of usually less than 1 per cent. Consequently, selection for floral characters which enable sufficient cross fertilization like more open flowering habit, duration of flower opening, improved anther extrusion in the male parent and stigma receptivity in the female parent is crucial for successful development of hybrids. The discovery of an effective cytoplasmic male sterility and pollen fertility restoration systems in wheat opened up new avenues for commercial hybrid seed production. Kihara (1951) pointed out the possibility of using male sterility by transferring the nucleus of common wheat into the cytoplasm of *Aegilops caudata*. Among the cytoplasms which in an interaction with *Triticum* nucleus bring about sterility, *Triticum timopheevii* seems to be most suitable for commercial production of hybrid seed. The economics of hybrid seed production is of major concern for successful hybrid technology. The contributing factors such as the plant population, male and female row ratios, plant spacings and input managements should be optimized for getting maximum hybrid seed at lower costs.

A Brief Historical Survey of F_1 Hybrids of Wheat

The research and development of hybrid wheat has a long history. Although the principal work has taken place during the 50 years beginning in 1950, the concept has been strongly influenced by developments in the production of pure line varieties of wheat over a longer period and by experience with hybrids of other species.

It is noteworthy that work on hybrid wheat has been carried out in majority of countries with significant cropping areas of this cereal. Major programmes of research have been run in many countries including Australia, Bulgaria, Canada, China, France, Germany, Hungary, India, Italy, Japan, Mexico, Netherlands, Pakistan, UK, USA, USSR and Yugoslavia. Hybrid wheat has been attractive to countries where wheat is an important crop. Production estimates for conventional wheat show that in 1990, the above mentioned 17 countries produced approximately 80% of the world's wheat. Although the view has been expressed that hybrid production, in general, appeals to commercial interests, in most of these countries, hybrid wheat research was carried out in the public sector while work in the commercial sector remained in second place.

Table 9.2. Chronology of events related to hybrid wheat development

Year		Land marks in heterosis breeding
1909	:	Hybrid maize first proposed by Shull
1912	:	The over-dominance theory of heterosis proposed by East and Hayes
1917	:	The dominance theory of heterosis proposed by Jones
1919	:	Heterosis was first reported in wheat by for plant height (Freeman)
1934	:	Heterosis first reported in wheat for yield (Engledow and Pal)
1951	:	Cytoplasmic male sterility introduced into wheat using *Aegilops caudata* cytoplasm (Kihara)
1953	:	Study the effect of maleic hydrazide as a chemical hybridizing agent (CHA) (published)
1957	:	USA is the first country to plan hybrid wheat production
1958	:	CMS research started on wheat in Kansas
1959	:	Nuclear male sterility reported in wheat (Pugsley and Oram)
1961	:	Fertility restorers found in adapted wheat varieties
1961	:	DeKalb Agricultural Association begins the first commercial hybrid wheat breeding programme
1961	:	The variety 'Gaines', the first semi-dwarf wheat released in the USA
1961	:	Fertility restorers found in adapted wheat varieties
1962	:	Source of CMS found in *Triticum timopheevi* (Wilson and Ross)
1962	:	First commercially feasible CMS system proposed
1966	:	McDaniel and Sarkissian proposed the theory of mitochondrial heterosis
1970	:	British Hybrid Cereals formed by Rothwell Plant Breeders and Rank Hovis McDougall to exploit hybrid wheat and barley
1971	:	DeVries commences publishing papers dealing with the suitability of wheat for cross-fertilization
1971	:	Rohm and Haas begin CHA research
1972	:	'XYZ' system proposed for the utilization of NMS (Driscoll)
1973	:	First feasible CHA discovered by Rohm and Haas
1974	:	First commercial CMS hybrid wheat released in USA
1975	:	Shell begins CHA research
1977	:	Experimental evidence adds further doubt to the theory of mitochondrial complementation
1980	:	The CHA WL 84811 discovered by Shell
1981	:	Hybrid wheat varieties released by Cargill in the USA and by DeKalb in Australia
1982	:	Monsanto starts HW program in US and Europe based on CHA genesis
1982	:	New CMS wheat hybrids make an impact in US market
1982	:	Renewed interest in hybrid wheat in Europe follows signs that Rohm and Haas and Shell CHAs are satisfactory
1982	:	Procedures for the official testing of hybrid wheat varieties published in France
1984	:	OECD begins work on international certification scheme for hybrid wheat
1984	:	Hybrid wheat varieties enter registration trials in the UK

Table 9.2. *Contd.*

Year		Land marks in heterosis breeding
1986	:	Hybrid wheat varieties released in Argentina by Cargill
1988	:	The CHA WL 84811 withdrawn after toxic residues found in F_1 seed
1990	:	Cargill ceases production and sale of hybrid wheat in the USA, but continues commercialization in Australia and Argentina
1995	:	ICAR (DWR) initiated work on hybrid wheat in a network mode through CHA and CMS approach
2000	:	Monsanto Co. stops GENESIS-based hybrid production and HW activities in US and Europe
2002	:	Dupont/Hybrinova stops Croisor-based hybrid production and HW activities in Europe

Extent of Heterosis

The basic requirements for hybrid development are commercially-viable seed production system and levels of heterosis. It is well known that many characteristics related to performance may exhibit heterosis. Performance in terms of grain yield is a result of a complex of contributory factors involving many loci. Stroike (1987) suggested seedling vigour, improved root system, disease and insect resistance, adaptability, increased yield and improved milling and baking characteristics as six possible factors to heterosis in wheat. It is possible for heterosis to be expressed by an F_1 hybrid in any part of the plant into which the products of photosynthesis are channelled. In this respect, it is necessary to consider the hybrid plant as a 'source' and 'sink' of assimilates. Virmani and Edwards (1983) have drawn attention to the need to balance these plant functions. Heterosis is unlikely to be directed exclusively towards factors associated with grain yield.

Commercial feasibility of a hybrid depends on the heterotic advantage over the best ruling variety of that agro-climatic zone. High heterosis over mid and/or better parent was reported in wheat by several workers. Heterosis was first reported in wheat by Freeman (1919) for plant height. Since then many workers have observed heterosis for various traits in wheat. Reports have provided ample evidence of significant positive heterosis over better parent for yield ranging from 0 to 100% in wheat (Briggle, 1963; Borghi and Perenzin, 1994; Larik *et al.*, 1995; Wang *et al.*, 1997, and Singh, 2003) but most of them are based on space planted and small plots. The heterosis over better parent for yield was noted to range from –40.62 to 29.75% among the winter × spring wheat (Chaudhary *et al.*, 1993). They reported that the higher yielding parents from different eco-geographical regions produced high heterosis and favoured hybridization between winter and spring wheats. Harvest index is an important indicator of source to sink relationship and therefore, Kumar and Ganguly (1993) advocated the possibility of increasing yield potential through better harvest index. The development of F_1 hybrid varieties of wheat was recently reviewed by Pickett and Galwey (1997). They also supported the hypothesis that heterosis in wheat arises from dominance with the additional factors of linkage and interaction of alleles. Under the field conditions, the minimum acceptable standard heterosis for yield is 20% for commercial exploitation of a hybrid. Significant standard heterosis of more than 20% was observed for yield and yield traits under drill sown conditions (Singh *et al.,* 1997, and Zehr *et al.,* 1997).

Yield components: Assuming that significant differences do not exist between hybrids and pure line varieties in plant establishment or resulting population levels, heterosis for grain yield must, of necessity, arise from an increase in the production of one or more of the plant's yield components. The weight of grain produced from a single plant will be the product of the number of fertile tillers per plant, the number of grains per ear and the weight of an individual grain. One of the principal underlying differences between the tillers and the number and weight of grains is the period of growth at which they are formed. As has been described by a number of workers including Sylvester-Bradley and Scott (1990), the establishment of potential tillers begins at the four leaf stage. Tiller fertility will then be determined at different times later on (Evans *et al*., 1972; Kirby and Appleyard, 1987; Whingwiri and Stern, 1982) although the effect on grain number will be decided early in floral development (Yu *et aI*., 1988). The timing of these phases will vary with variety (Rawson, 1970; Kirby and Appleyard, 1987). On the other hand, grain weight is largely determined post-anthesis. The number of grains per ear is, of course, the product of number of spikelets per ear and number of grains per spikelet. As already mentioned, many studies on hybrid wheat were undertaken on spaced plants. There has been a lack of correlation with the yield of normal density plantings. It is likely that yield components contributed in differing degrees to the lack of agreement between spaced plantings and normal cropping.

The expression of any potential heterosis for a component is likely to be strongly influenced by environmental factors which may affect genotypes differently. However, many researchers have attempted to determine the component of yield responsible for heterosis in hybrid wheat. The principal studies on the source of grain heterosis indicated that the tillers, grain number and grain weight as major components for realizing high yield potential in winter wheats as well as spring wheat germplasm. The results indicate that a single component is not the sole source of heterosis. This is not surprising since it is well known that there is considerable variation between pure line varieties of wheat in the performance of individual components. The inheritance of each of the three components of yield is highly complex with the involvement of many genes which may show different modes of inheritance. Part of this complexity arises from negative compensation between components as has been shown by Hsu and Walton (1970) and Pinthus and Millet (1978). It has also been found that poor seed set may be associated with low assimilate levels. Although the studies lack complete agreement, the general trend in these polygenic characteristics are inherited in an additive manner with partial or complete dominance and frequent interaction. The types of gene action noted by a number of investigators are summarised in Table 9.3.

Table 9.3. Studies of the mode of inheritance of three yield components

Author	Method	Tillers	Grain number	Grain weight	Yield
Whitehouse *et al.* (1958)	Wr Vr regress.	A, D	A, D	A, D	NA, I
Lupton (1961)	Wr Vr regress.	R	D, I	A, D	D, I
Hsu and Walton (1970)	Wr Vr regress.	A, D	A, D	A, D	A, D
Halloran (1975a,b)	Wr Vr regress.	D, I	A, R	A, D	D, I
Fedin (1976)	Wr Vr regress.	A, D, I	D	A, D, I	A, D, I
Snape (1982)	TTC, JST, LIE	D, S	D, I	A, D, I	D, S

Key: A = additive, D = dominant, R = recessive, I = non-allelic interaction, NA = non- additive, S = no interaction; TTC = triple test cross, JST = joint scaling test, LIE = linked interactive effects.

The postulated benefits for hybrids have been considered from theoretical principles of the nature of heterosis and the mode of inheritance of the important genes. The findings reviewed support the belief that heterosis arises as a result of the bringing together of dispersed dominant genes and non-allelic interaction. The question is complicated by additive genes and the fact that many characters are polygenically controlled, making the behaviour of the F_1 difficult to predict for many characters.

The contention that heterozygosity *per se* imparts increased vigour has scarcely any support from the very considerable experimental evidence accumulated in the literature. Moreover, the analogous belief that heterosis occurs at the mitochondrial level in wheat has not been confirmed by investigations or by advances in molecular genetics. Trial results may be divided into two groups on the basis of the method of investigation and the nature of the material worked on. Thus, the small-scale, commercially unrepresentative trials of hybrids from crosses of relatively 'primitive' wheat varieties gave some very promising results. In contrast, the extensive work using modern varieties as parents has generally demonstrated no more than a small advantage for hybrids over the best pure line varieties, with an increase of more than 5% difficult to achieve and sustain. Together with the theoretical considerations, this gives good support to the hypothesis that heterosis may be fixed.

Parental Diversification for Yield Components

As the grain yield from a single plant is considered to be the product of the number of fertile tillers per plant, the number of grains per ear and the weight of an individual grain, there is need to have parental lines with better yield components that can be accumulated for harnessing heterosis at commercial level. Such parental lines can be developed by pre-breeding or diversification through utilization of diverse germplasm lines.

Hermsen suggested pre-breeding as an integrated breeding approach in potato for the use of quantitative traits in distant species and defined it as a process in which the traits scattered in different accessions of a species are combined at one source. In order to widen the genetic base of bread wheat, the emphasis has been laid on introgressing genes from unexploited *buitre* types, synthetic hexaploids and Chinese sub-compactoid ear germplasm. The buitre lines developed by Rodriguez Ricardo have robust stem, long spikes, more spikelets, more grains per spike, large leaf area and broad leaves. The synthetic hexaploids developed by Dr. Mujeeb Kazi at CIMMYT were endowed with genetic richness for high grain weight, delayed senescence (stay green), HMW glutenins, resistance to KB and yellow rust. Similarly Chinese germplasm lines have robust stem, more grains/spike and new sources of yellow rust resistance. However, these elite germplasm obtained had certain inherent limitations such as susceptibility to rust diseases and seed shriveling in *buitre* types, hard threshing, red grain colour, delayed heading and maturity in synthetic hexaploids and very soft grains (prone to insect damage) in the Chinese material. These unique germplasms were evaluated for possible use in hybrid program as diverse gene pools and promising lines were used in breeding programme. The desirable attributes from buitre types, synthetic hexaploids and Chinese germplasm were introgressed into PbW 343 and WH 542 background. The advanced bulks developed through utilization of diverse matreial have shown wide range of variability. The introgression for 1000-grain weight (HT lines) was also observed from the Chinese germplasm lines and a number of transgressive segregants were obtained having 1000-grain weight of more than 65 g. These lines also showed a wide range for various yield component traits (Singh, 2007).

Floral Biology

Floral structure, anthesis and anther dehiscence patterns in wheat make this crop strictly autogamous (Percival, 1921; Leighty and Sando, 1924). De Vries (1971) reviewed the floral biology of wheat keeping in view the possibility of hybrid seed production. For hybrid development, adequate cross pollination attributes are the necessity. The occurrence of outcrossing in wheat and the tendency for some cultivars to outcross more than others was well established (Griffin, 1987). Around six decades ago, Leighty and

Taylor (1927) demonstrated that outcrossing occurred more frequently in late emerging spikes of wheat cultivars. The extent of natural outcrossing in cultivated varieties of wheat was observed to be 0.0 to 6.05 per cent (Heyne and Smith, 1967; Martin, 1990; Hucl, 1996; Singh, 2006). A positive correlation of outcrossing with the length of flowering period was observed (Martin, 1990). Mixed chasmogamous/cleistogamous type flowering and autogamous/allogamous mode of pollination in wheat was observed (Chelak, 1989; Chhabra and Sethi, 1991).

Important floral traits that influence outcrossing in wheat are stigma size, anther size, anther extrusion, pollen number and pollen viability. The stigma length of wheat genotypes has been noted to be 2.13–5.2 mm (Percival, 1921; Komaki and Tsunewaki, 1981; Singh, 2005). Anther length has also been investigated by a number of workers and range was observed from 3.0 mm to 5.09 mm. Anther extrusion, an important feature in wheat, has been observed from 14.1 to 93.0% (Singh and Joshi, 2003). Phenotypic differences among wheat cultivars for days to heading and anthesis, anther extrusion, anther size, pollen grain size and pollen viability, stigma length, duration of floral opening and openness of florets were also observed by Chowdhary *et al.* (1994), Singh and Joshi (2003) and Singh *et al.* (2007). They also suggested that the selection for long anthers, high rate of anther extrusion and more openness of florets may be effective in promoting natural cross pollination. The pollen viability in wheat ranged from 81–98.6% (Hucl, 1996; Singh and Singh, 2001). In general, flowers of wheat reach their peak opening between 9.00 and 11.30 a.m. on a sunny day. Subsequently, a second flush of flower opening occurs between 15.00–17.00h (Virmani and Edwards, 1983). The largest separation angle between the glumes of the first two florets of spikelet was observed to be 16–40°. Wheat florets get closed within 12–20 minutes of flower opening, the extreme values being 8 and 39 minutes (Singh and Singh, 2004). Gorin (1968) observed 80–90 per cent open florets in most of the bread wheats. Flowering continues throughout the day, with 2–6 days required for a spike to finish blooming. After the first round of blooming, unfertilized florets open again, exposing the stigma for another 2–3 days and the ovaries remain receptive for an additional 2–3 days (Hoshikawa, 1960). Likewise, in male sterile plants the second round of flowering starts 36 hours after the first round and florets remain open for another 5 days (Molnar-Lang *et al.*, 1980).

There is a positive association between anther size and the quantity of pollen produced per anther (Beri and Anand, 1971). Further, within wheat cultivars, a large variation is reported for the number of pollen grains produced per anther. Under optimal field conditions (20°C, 60% RH), wheat pollen is viable for approximately half-an-hour (D' Souza, 1970; De Vries, 1971). All the viable pollen grains that have access to the stigma surface germinate as soon as they come in contact with the stigma (Chandra and Bhatnagar, 1974). Because the entire length of stigma is receptive to pollen grains in wheat, the location of pollen landing on stigma, does not affect pollen germination. Stigmas are receptive for 2 to 13 days after anthesis, and are most receptive for the first 2–5 days (De Vries, 1971). Outcrossing may be increased by a larger stigmatic surface and longer style exertion. The angle of separation between the lemma and palea is important in such a way that both branches of the stigma protrude from the floret during flowering. Wheat pollen is relatively heavy compared with pollen of other grasses and travels up to 1 m from the source of pollen (Lelley, 1966).

Male Sterility System in Wheat

Wheat is monoecious and therefore a line designated as female must not be allowed to produce pollen capable of fertilisation whilst acting as a parent. It must, however, be able to do so for the maintenance of the line. Since wheat flowers are hermaphrodite, the labour requirement for mechanical emasculation is clearly prohibitive. Methods for procuring male sterility in plants may be divided into two groups on the basis of the role of the genotype. Systems involving the plant's genes include cytoplasmic male sterility, nuclear male sterility and self-incompatibility (Wricke, 1989). On the other hand, there is no direct genetic involvement in chemical hybridization or hand emasculation. Three methods for obtaining male sterility have been proposed for wheat: these utilise, respectively, cytoplasmic and nuclear genes, nuclear genes, or chemicals, but CMS and CHA approaches are widely used in wheat as pollination control system.

Cytoplasmic genetic male sterility system: This system is commonly known as CMS system and involves three types of parental lines namely, the male sterile line (A-line), the maintainer line (B-line) and the fertility restorer line (R-line). Hybrids are the resultant of A × R crosses. Although the CMS approach has advantages of complete sterility in female lines, absence of toxicity and chemical registration costs but it requires much pre-breeding, long term commitment of resources and labor and is a slow process. Cytoplasmic-genetic male sterility is visualized as an essential genetic tool to produce F_1 hybrids in self-pollinated crops.

Sources of cytoplasmic-genetic male sterility: Kihara (1951) reported the cytoplasmically induced male sterility in wheat for the first time by transferring the common wheat genome into *Aegilops caudata* cytoplasm. The possibility of hybrid wheat became apparent only when Wilson and Ross (1962) discovered another effective source of male sterility in *T. timopheevii* 'Bison' cross. Following the report of Kihara (1951) on *Ae. caudata*, Curtis and Johnston (1969) also observed *Ae. caudata* and *Ae. ovata* as good sources of cytoplasmic male sterility. A number of species from genera *Triticum* and *Aegilops* were reported as source of male sterility but *Triticum dicoccoides* var. *spontaneovillosum*, *Ae. aucherii* (Stehno and Apltaverova, 1980, 1981); T. *zhukovskii* (Apltaverova *et al.,* 1981);

Ae. ventricosa (Zhang and Yang, 1989); and *Ae. kotschyi* (Ushiyama *et al.*, 1996) were considered as a good sources of cytoplasmic-genetic male sterility. *Triticum timopheevii* was found to be a major source of male sterility in wheat by several workers (Curtis and Johnston, 1969; Apltaverova *et al.,* 1981; Il'ina, 1985, and Ushiyama *et al.,* 1996) and is now the most widely used cytoplasm in hybrid wheat production.

Fertility restoration system: *T. timopheevii* and its derivatives were also observed as potential source of fertility restoration system in wheat (Schmidt *et al.*, 1962; Wilson, 1972; Panayotov *et al.,* 1986). Other sources of fertility restoration in wheat were wheat varieties Primpei (Oehler and Ingold, 1966), Lal Bahadur, Ridley and HD1944 (Karim and Singh, 1984), and related hexaploid species *T. dicoccoides* var. *spontaneovillosum* (Panayotov *et al.,* 1986) and *T. spelta* var. *duhamelianum* (Kihara and Tsunewaki, 1964).

Genetics of male sterility-fertility system: Two incomplete dominant genes with epistatic action were reported to control fertility-sterility system in wheat by Nettevich and Naumov (1970). A gene *ms2* for male sterility was observed on short arm of chromosome 4D by Liu and Yang (1994). Liu *et al.* (1996) reported that the male sterility was controlled by double recessive nuclear type action and suggested involvement of up to 2 genes, designated as *fms1* and *fms2* for male sterility. A single recessive gene was also found responsible for male sterility (Il'ina, 1984; Chhuneja and Minocha, 1993).

Fertility restoration in wheat has been reported to be under the control of monogenic, digenic as well as polygenic control. The complexity of the genetics of restoration is evident from the fact that restorer genes are carried on many chromosomes in wheat (Sage, 1976). These include chromosomes 1, 2, 5, 6 and 7.

Three genes, located on chromosome 1A, 6B and 7D of *Triticum timopheevii* were found responsible for complete fertility restoration (Wilson, 1972). However, Ralko (1982) observed all the three (monogenic, digenic and polygenic) mode of inheritance. Chromosome 1B of Chinese Spring was also identified as carrier of fertility restoration gene (Tsunewaki, 1982; Tsujimoto and Tsunewaki, 1984). El-Kadi *et al.* (1983) observed 2–3 major genes in R line for fertility restoration while two dominant (Gotsova *et al.*, 1987) and two additive genes (Wrobel, 1989) were found responsible for fertility restoration. Multiple recessive genes with minor effects were responsible for control of fertility restoration in wheat (Murai *et al.*, 1995). Ma *et al.* (1991) observed that at least one gene is responsible for controlling fertility restoration in wheat. On the other hand, Du *et al.* (1991) reported that two genes, responsible for fertility restoration (*Rf1* and *Rf4*), were located on chromosomes 1A and 6B, respectively. Ikeguchi *et al.* (1994) located the fertility restoration gene *Rfv1* on chromosome 1B.

Cytoplasmic male sterility and male fertility restoration systems in hexaploid wheat are conditioned by interaction involving dosage of *Rf* genes (restoration of fertility) and *Fi* genes (fertility inhibiting) in polyploid nucleus and cytoplasmic genes of the related alloplasmic species. The plant vigour restoring genes *Vi* also play important role. The functions of *Rf, Fi* and *Vi* were not mutually exclusive and the same genes may have produced different genotypes in combination with different cytoplasms and wheat nuclear genes (Mann, 1992). More than 10 different *Rf* genes have been reported (Kucera 1982; McIntosh and Cusick 1987: Huang 1990: Chen *et al.*, 1998). The adequate fertility restoration of R-lines to T-type CMS needs more than two doses of these Rf genes. Stable, complete fertility restoration needs even more (Huang, 1990; Chen *et al.*, 1998). Although fully self-fertile alloplasmic lines with homozygous Rf genes are easily obtained, most of the hybrids with the CMS lines are not fully fertile. The problem also remains in other alloplasmic wheats such as the K-type (with cytoplasm of *Ae. kotschyi*) and the V-type (with cytoplasm of *Ae. ventricosa*) hybrids. To improve the male fertility restoration of common wheat for *T. timopheevii* cytoplasm, Chen (2003) proposed the new model of A-line/R*-line//R-line in the production of hybrid wheat.

Effect of cytoplasms: Different sources of male sterility system carry their own negative effects on growth and productivity of wheat. The cytoplasm from *Triticum timopheevii* has been found to exhibit tendency towards pistiloidy and meiotic instability, reduced dry matter weight, kernel shriveling, low seed germinability, higher anther extrusion, increased tiller production and ear length, lower grain weight/ear, and reduced grain filling by several workers. The *Aegilopes* cytoplasms were also found to cause delayed heading, reduced plant height, longer ears, more productive tillers/plant and shrivelled grains when crossed to maintainer/restorer line owing to late maturity. Zhang and Yang (1989) found that *Ae. ventricosa* cytoplasm gave rise to stable sterility and normal seed plumpness of male sterile F_1s and F_2s and did not result in harmful maternal effects.

Breeding scheme for F_1 hybrids using CMS: The CMS system for producing hybrid seed has been described many times (Rodriguez *et al.*, 1967; Sneep *et al.*, 1979; Wilson, 1984 and Kaul, 1988). In this system, the production of a single cross hybrid requires three lines, two to provide a female parent and one for the male. On the female side, after a genotype has been selected for combining ability, the first requirement will be to transfer it to a sterilising cytoplasm. In most species, CMS is incorporated into a line of the female parent by backcrossing using the line with the male sterile cytoplasm as the recurrent parent and the normal, male fertile line, as the donor parent (Newton, 1988). The CMS line is, by convention designated the 'A' line. Workers vary on the subject of the number of backcrosses required to produce a CMS line: Zeven (1967) states six (taking three years); Macer (1972) eight, Tsunewaki *et al.* (1980) ten, Bingham and Lupton (1987) five back cross cycles. Although it is possible for the CMS lines, once developed, to remain true to type, problems are sometimes experienced in the maintenance of sterility. Wilson (1968b), for example, confirming that male-sterile 'Bison' had been multiplied successfully through 14 generations, noted some exceptions in other lines.

A line carrying the same nuclear genotype in a fertile cytoplasm, usually from *T. aestivum,* is used for pollinating the A line for maintenance purposes. This line is known as the 'B' line. The preservation of CMS relies on the non-transmission of the cytoplasm by the pollen of the male. The male parent, in addition to being selected for combining ability, must normally carry genes to restore fertility to the F_1 hybrid. Restorer genes are incorporated into the male parent by means of a crossing programme, which may also include backcrossing. Restorer genes are given the designation '*Rf*' and it is customary to denote the restorer as the 'R' line. In wheat, the B and R lines are both maintained as normal inbred lines, without difficulty, using conventional seed production arrangements. The restorer locus or loci will be heterozygous in the F_I hybrid. Segregation will, therefore, take place in subsequent generations, precluding the use of CMS in the production of F_2 hybrids of wheat. The following diagram describes the basic system for producing an F_1 hybrid by means of CMS. This uses the commonly adopted notation whereby the CMS is termed the 'A line', its maintainer with the same nuclear genotype is called the 'B line', and the restorer acting as the male parent of the F_1 hybrid is the 'R line'.

PRODUCTION OF AN F_1 HYBRID VARIETY USING CMS

Chemically Induced Male Sterility

The limitations of cytoplasmic male sterility system, *viz.*, unstable nature, undesirable linkages and need for use of maintainers have prompted breeders to develop simple and more efficient methods to create male sterility by other means like use of chemicals, mutagens for induction of male sterility. Male sterility induced by chemical hybridizing agents is relatively more convenient to use because there is no need to maintain, it does not require any pre-breeding, fast and relatively easy to implement and, virtually any combination can be explored. Besides, it has disadvantages of being very expensive to develop and register, highly crop growth stage-dependent and weather-dependent, often incomplete sterility in female and somewhat phytotoxic effects. Compared with CMS systems, an effective CHA allows the production of large numbers of parental combinations and permits the evaluation of a number of inbreds for combining ability and/or breeding value. This substantially reduces the time required for hybrid development. More than 40 chemicals have been patented as potential CHAs world over, out of which etherel and maleic hydrazide are most commonly used. Complete male sterility by spraying of etherel at early stem extension stage was observed along with reduction in plant height due to reduced internode length, plant development, yield and increased number of fertile tillers. The Shell compound WL84811 and the Monsanto CHA "GENESIS" are well known current generation CHAs for inducing male sterility. The appropriate stage for higher efficacy of CHA was observed at spike length of 7–8 mm to early boot stage and the late sown crop was found more responsible to CHA than normal sown crop. The hybrid development process via CHA route involves the induction of male sterility in female lines through chemical spray at appropriate stage and the pollination for seed setting.

Attributes of an ideal CHA have been suggested by several authors including Liable (1974), Virmani and Edwards (1983) and Pickett (1993). A collective list of those attributes for an ideal CHA includes:

- Selectively induces only male sterility with no effect on female fertility
- Produces male steriles that are readily apparent
- No phytotoxic effect on the treated parent
- Effective on all genotypes of a species
- Effective in a wide range of environments
- Has systemic activity or persistence to sterilize early and late tillers or plants in the field population
- Has flexibility in time of application to overcome adverse weather effects and permit treatment of large areas
- Has considerable dosage flexibility to permit a safety margin for application
- Achieves sterility with a single treatment
- Does not adversely affect F_1 seed quality, or F_1 seedling or plant vigour
- Effective on several genera
- Economical to synthesize and practical to apply
- Safe and non-toxic to the environment.

While there are a rather large number of chemicals that cause male sterility in wheat, there are very few that meet most of the above requirements. Once a chemical is identified as having some CHA activity, much work needs to be done to identify the optimum time and rate of application. Additionally, application rates can vary with environment and genotype and can be modified with the use of chemical surfactants.

Genetic Male Sterility

A number of genes have been identified which prevent the production of pollen capable of fertilisation. It is possible for male sterility to be achieved without a significant reduction of female fertility. Since most normal wheat varieties will act as restorers of fertility, necessary for the F_1 plant, it is only necessary to breed special female parents. Genes controlling nuclear male sterility fall into two categories: those showing dominance or near dominance for sterility and those which are recessive. Fertility in plants is usually encoded by dominant genes. Mutations of a single gene are the probable cause of the majority of examples of nuclear/genic male sterility in plants. These may cause an alteration of a gene to the recessive state leading to loss of function and sterility. Except in special circumstances, recessive genes for male sterility are highly desirable because they usually facilitate identification of homozygous sterile progenies for F_1 production. However, it is unlikely that all the NMS genes act in the same way. Some, for example, act as dominants. There have been many reports of genetic male sterility (GMS) of wheat in the literature. Five GMS loci have been located to chromosomes so far. These loci are:

1. *ms1,* a recessive gene located to chromosome arm 4BS;
2. *Ms2* and *Ms3*, two dominant genes located to chromosome arms 4DS and 5AS respectively;
3. *Ms4*, a dominant gene located to chromosome arm 4DS, and
4. *ms5*, a recessive gene located to chromosome arm 3AL.

Systems of Genetic Male Sterility in Wheat

The big obstacle in the use of NMS systems is maintenance of the male sterile parent. Normally this is achieved within a heterozygous population (Hermsen, 1965) but this conflict with the requirement that the female parent must not segregate for fertility in the hybrid seed production field. In principle, this applies for all species in which male sterile seed is produced by GMS; the problem is particularly serious for wheat because low multiplication rates make hand roguing of fertile segregants uneconomic and unsatisfactory. Maintenance requirements have, therefore, been the main consideration in the proposed systems for NMS production in wheat.

The 'XYZ' system: This system was designed to allow maintenance of a GMS line in which none of the 42 normal chromosomes carried a factor for fertility. Fertility to produce the female parent seed would be obtained from another line: identical except that it carried a chromosome addition encoding fertility. The XYZ system of producing hybrid wheat (Driscoll, 1972) relies on the addition of chromosome 5R derived from *Secale cereale* L. to the male-sterile Cornerstone mutant (Driscoll, 1977). The necessary fertility for maintaining the line would come from a corresponding fertility factor on a homoeologous alien chromosome addition line. A requirement of the XYZ system is that the alien chromosome should not pair with any from the normal complement. Since the dominant marker character-hairy peduncle associated with 5R is an adult plant trait, differentiating between male-sterile plants without the marker and the male fertile plants having the marker trait will take place at a post heading plant developmental stage in the field. So it is not a practical system, as it is difficult to rogue out the male fertile plants in the short time between heading and pollen shedding. A more efficient marker character associated with an alien chromosome is needed for logically modifying the XYZ system.

The 'VE' system: The line 'VE' was obtained from a cross of *Triticum vulgare* and *Elytrigia elongata,* the designation 'VE' being derived from the first letters of the specific and generic names of the parents. When the line was grown in populations heterozygous for the dominantly controlled male sterility, the progeny is approximately 75% sterile (Shaowen and Shan, 1980). This is used for bulk maintenance of the line. When male sterile seed is required, the segregants having normal fertility are first rogued out from the population. The remaining plants, although largely sterile, produce enough pollen to allow selfing (Driscoll, 1986). However, as a consequence of incomplete male sterility in the female parent seed, it is likely that the genetic purity of the hybrid produced will be rather low.

***Genic male sterility associated with chromosome 4A*:** Four male sterile lines namely 'Pugsley', 'Sirocco', 'Cornerstone' and 'Probus' developed by different workers, apparently from mutations involving a single locus on chromosome 4A. As mentioned below, two of these were spontaneous mutants (Pugsley and Sirocco) and two were produced by ionizing radiation (Cornerstone and Probus).

***Pugsley's male sterile line*:** Pugsley and Oram (1959) in Australia identified this recessive male sterile line as a spontaneous mutant in an F_3 progeny of a cross Kenya Farmer/Bearded Javelin 48. Male sterility was transferred from this line to 'Chancellor' by Briggle (1970) and has compatible genetic background for must expression governed by this gene. After a strict selection programme, high level of sterility was achieved although small, infertile pollen grains were produced.

***Sirocco male sterile line*:** This male sterile line was developed at Cambridge from a spontaneous mutant in a cross of commercialized varieties (Lupton *et al.,* 1980).

***Cornerstone male sterile line*:** Driscoll (1977, 1979) in Australia developed the mutant line Cornerstone by irradiation. Irradiation of euploid pollen at about the time of pollen shedding and its application to monosomics 4A produced one mst mutant, the Cornerstone mutant. This gene was found to be unaffected by environment or genotype.

***Probus male sterile line*:** The normal fertile winter wheat variety Probus was subjected to X-ray irradiation of seeds by Fossati and Ingold (1970) in Switzerland. The probus male sterile line segregating for male sterility was recovered from the progeny. The sterility is complete, as its anthers have empty pollen grains and its florets remain open at flowering.

Other Genes Conferring Male Sterility

S738: Athwal *et al.* (1967) working in India reported a specific type of male sterility caused by the interaction of three recessive genes having an additive effect which occurred in F_2 population of a composite cross received from Mexico. The degree of male sterility depends on the number of ms genes; the lesser the number; the lesser is the expression of male sterility. Thus, the genotypes having 1, 2, 3 *ms* genes exhibit 36–49%, 47–68% and 73–97% male sterility, respectively. This expression was further influenced by environment and by modifying genes.

Taigu: In Chinese bread wheat, a dominant Ms gene induces mst. These plants segregated from the progeny derived from a composite hybridization with eight materials of *T. aestivum* from Taigu County in the Shangai Province of China. The sterility was found to be associated with aneuploidy. Only plants lacking chromosome 4D were sterile. This gene was designated as Ms_2. Bing-Hua and Jing-Yang (1986) working in China reported that Taigu had a potential use in breeding work especially in wide crossing.

UC9109-9: UC 9109-9 thought to be of mutant origin, was identified in the progeny of dwarf × normal height cross in the USA (Jan and Qualset, 1977). Three recessive genes controlled this source. It resulted in a range of expression of male sterility which varied from very low levels of seed set when selfed to the transformation of anthers to pistil-like structures.

'VE': The line 'VE' was obtained from a cross of *Triticum vulgare* and *Elytrigia elongata*, the designation 'VE' being derived from the first letters of the specific and generic names of the parents. Shaowen and Shan (1980) reported that when grown in isolation, the line produced around 75% male sterile plants. Driscoll (1986) has shown that the line contains a pair of *E. elongata* chromosomes that carry the fertile recessive allele to the male sterile gene accounting for the 25% fertile plants.

'Yuma' x 'Capeiti': The durum cross 'Yuma x Capeiti' is of interest because the breeders, Bozzini and Scarascia-Mugnozza (1968) working in Italy, felt that a transfer of male sterility to *T. aestivum* was theoretically possible. In the progeny of the durum cross, male sterile individuals were found in which sterility was controlled by a single recessive gene.

Advantages and Disadvantages of GMS

Wilson (1984) noted that, despite prolonged investigations, GMS had not been utilised commercially for cereal crops. Apart from use in China, there is little to suggest that GMS has reached commercial use. The basic requirement for the GMS system is the existence of genes encoding male sterility. The drawback with most systems is large scale production of uniformly sterile seed. There is much work to be done before this method of producing male sterility could be used. GMS, however, offers certain potential benefits including the absence of the need for special restorer varieties and reasonable freedom from the damaging side-effects apparent in many forms of CMS. Most conventional varieties of wheat may be used for this purpose, making it possible to select male hybrid parents from a wide range of varieties. The disadvantage of GMS wheat was the maintenance of the male sterile line. In general, this requires production of a population heterozygous at the locus controlling fertility. The fertile segregants must then be removed from the hybrid seed production field prior to flowering. Although a marker gene linking to a conspicuous morphological character to fertility could, in technical terms, allow removal of the male fertile plants, in practice, the scale of operations would preclude this approach. It is possible that new techniques could find application in developing GMS. For example, if it becomes possible to manipulate cereal genes directly, it may then be possible to complete the assembly of a workable system. It may, however, be more worthwhile to attempt to link a gene for herbicide tolerance to male sterility to provide the means to remove fertile segregants in an NMS line.

Environment Sensitive Male Sterility (EMS)

Photoperiod-sensitive cytoplasmic male sterility (PCMS): Photoperiod-sensitive cytoplasmic male sterility caused by an interaction between *Aegilops crassa* cytoplasm and *Triticum aestivum* cv. Norin 26 nucleus was first reported by Murai and Tsunewaki (1995). Alloplasmic 'Norin 26' shows almost complete male sterility under long-day conditions of 15 hours or longer, but male fertility under short-day conditions of 14.5 hours or less. Japanese wheat cultivars were classified into two groups based on the effect of this cytoplasm, *i.e.* PCMS group (-Rf gene) and the fertile group (+Rf gene). The PCMS is maintained and multiplied by self-pollination under short-day conditions (*e.g.* natural day length in autumn sown conditions) and hybrid seeds can be produced through out-crossing of a PCMS line with a pollen parent under long-day conditions (*e.g.* natural day length in spring sown conditions). In contrast to the CGMS system, this system requires only a PCMS line and a pollen parent (restorer line). Moreover, studies on evaluation of F_1 hybrids obtained by hand pollination between the PCMS lines with restorer cultivars revealed the other advantages, namely, no undesirable influence of the *Ae. crassa* cytoplasm on the agronomic characters in the F_1 hybrids, and a sufficient level of fertility restoration in the F_1 hybrids (Murai, 1977a). Two kinds of fertility restoration systems against the PCMS are known. One is involved with a multiple fertility-restoring (Rf) genes in wheat cultivar 'Norin 61' located on (at least) chromosomes 4A, 1D, 3D and 5D (Murai, 1997b). The other is controlled by a single dominant major Rf gene *rfd1*, located on the long arm of chromosome 7B in the wheat cultivar "Chinese spring (Murai and Tsunewaki, 1994). Murai (2002) found that fertility restoration ability of a set of multiple Rf genes in 'Norin 61' was higher than that of the *Rfd1* gene in 'Chinese spring".

Photoperiod sensitive genic male sterility (PGMS): Fisher (1972) observed that subjecting wheat plants to 10h of photoperiod at a stage when stamen primordia were visible on the first floret of the most advanced spikelet of the main shoot led to the transformation of stamens in to ovaries. Ovules were found to develop on the anther lobes to make plants sterile. Zhou *et al.* (1997) and Luo *et al.* (1998) used photoperiod sensitive ecotype male sterile (EMS) wheat lines ES-3, ES-4, ES-5, ES-8 and ES-10 for developing two-line wheat hybrids.

Temperature-sensitive genic male sterility (TGMS): Certain wheat varieties carry genes, which render male fertility sensitive to environmental conditions such as reduced temperature (Qian *et al.*, 1986). In regions having consistent climates, it may be possible to utilize such effects for production of F_1 hybrid seed for use in crop production in different environment. Jan (1974) reported thermo sensitivity in wheat for producing genic male sterility.

Photo thermo-sensitive genic male sterility (PTGMS): Long photoperiods or higher temperatures or both enhance pollen sterility. Thus, with increasing latitudes toward the north, photoperiod increases that induce male sterility. Chinese scientists at Hunan Agricultural University also reported PTGMS in wheat in 1992 (cf. Yuan, 1997) and developed the two-line wheat hybrid E-you 137 which showed superiority at a low seeding rate (45 kg/ha) by yielding 20% more than conventional varieties of wheat (cf. Yuan, 1997). Cytoplasmic-genic male sterility in wheat with *Aegilops umbellulata* or *Secale cereale* cytoplasm was stable in autumn and unstable in spring when grown in the glasshouse (Mann, 1973), indicating photo-thermo sensitivity.

Micronutrient deficiency-induced male sterility: Deficiencies of copper (Agarwala *et al.,* 1980), boron (Rerkasem and Jamjod, 1997) and some other micronutrients are reported to induce male sterility in wheat. High phenotypic variation has been reported in sensitivity to the deficiency of these micronutrients. Very sensitive types are completely male sterile under micronutrient deficiency conditions. The partial male sterility induced by insufficient copper nutrition in wheat resulted in delays in ear, anther, or pollen grain development. Under severe deficiency, the flowers did not open and the anthers did not contain pollen or were indehiscent; the pollen grains formed were small and had low germinability. Jewell *et al.* (1988) reported that pollen sterility induced by copper deficiency in barley and wheat was related to tapetum dysfunctioning characterized by cell hypertrophy. Azouaou and Souvre(1993) found that copper deficiency induced nearly complete sterility of the pollen formed and inhibited grain production. Rerkasem and Jamjod (1997) suggested the boron method of hybridization in wheat, in which boron deficiency was used as a selective medium or fertility and male sterile female parents and fertile male parents were provided by genotypic variation in response to low boron content. The female parents can be maintained simply in soils with sufficient boron content, unlike genetic male sterility, which can be maintained only through heterozygous populations. The chances for successful development of the method, however, could be strengthened by a better understanding of the physiological, genetic, and molecular bases for genotypic variation in response to low boron content. Further, the basic understanding may in turn lead to better insights into the nature of reproductive development, and therefore opportunities for more effective control in wheat and other smaller grains.

COMMERCIAL HYBRID SEED PRODUCTION

The most important requirements of cross-fertilization and hybrid seed production are:

- Flowers must open at peak flowering period
- The stigma must be receptive to accept the pollen from the male parent
- Anthers must release pollen outside the flower by good anther extrusion
- Slight wind must be present to carry the heavy pollen from male to female plants
- Pollen must be viable till fertilization.

Two constraints apply to hybrid seed production in wheat. First, the multiplication rate of the species is naturally low. Second, as cross-fertilization may frequently be inefficient, leading to low levels of seed set. In the hybrid seed production field, it is necessary to maintain careful separation of male and female parents. If this is not achieved, the hybrid seed will be contaminated with parental lines and unwanted hybrids. In view of the poor transportation of pollen, the distance of separation of parents must, however, be the minimum consistent with the cost-effective use of farm machinery. Where a CHA is used, it is, of course, necessary to perform an accurate application of the chemical without leaving the female parent male fertile or treating the male parent. Amongst the more important characteristics benefiting hybrid seed production in wheat, a consistent climate with warm and sunny conditions without excessive rainfall is needed. However, excessive heat may be harmful. If production is to be carried out away from the area of use, parental varieties will have to be selected to synchronize in the production environment. It is also apparent that the environment within an area can, to a degree, be optimized for wind pollination by choice of a field exposed to wind and by aligning the beds in such a way that the prevailing wind can move pollen from the male parent to the female.

Investigations concerning male: Female ratio for hybrid seed production of wheat have demonstrated yield advantage at variable ratios. Wilson (1968) reported 70% seed set in 1:1 ratio of the sterile and pollinator parents but suggested the possibility of profitable management of 2:1 ratio. Maximum yield at 1:1 ratio of male sterile/pollinator parent followed by 2:1 and 3:1 ratio in production/hectare basis was observed by Miller and Lucken (1976). However, female : male row ratio of 2:1 or 3:1 was also reported for getting maximum seed set (Singh and Singh, 2006). Araki (1990) observed 0–80% grain set in the male sterile line and reported higher seed setting when the male sterile line headed 5–10 days before the restorer lines. He also observed a close relation of seed set with the extent of synchrony in heading of maternal and pollen parent. A direct effect of cross-fertilization on seed yield and the cost of hybrid seed were also reported. The yield of hybrid seed can be computed by

$$\text{Yield of hybrid seed } (F_1) = a\,(b/100)\,(c/100)$$

where, a is normal yield of wheat (t/ha), b is area of female parent (%), and c is seed set (%).

Hybrid Wheat Research at CIMMYT

The work on the development of hybrid wheat was started in 1962 at global level in many countries. Ing. Riccardo Rodriguez initiated the research efforts at CIMMYT in 1962. The elite CIMMYT lines were transferred with *T. timopheevii* cytoplasm, the fertility restorer (Rf) genetic stocks were developed and the experimental hybrids were produced and evaluated but with the advent of semi-dwarf high yielding wheat varieties, the emphasis was diverted only to popularization and genetic improvement of pure line varieties and, as a result, the research efforts on hybrid wheat got distracted. The work was discontinued as no significant results of heterosis were observed for commercial exploitation.

The research efforts were readdressed at CIMMYT during 1997–2002 in collaboration with the Monsanto Co. with the objectives to develop a practical hybrid wheat production scheme in northern Mexico, based on the GENESIS (CHA) technology and to identify spring hybrid bread wheats with superior yield potential, leaf rust resistance and acceptable quality under optimal conditions. In these efforts, the doses and crop stage were optimized for complete male sterility, the female sterility was monitored and adequate levels of male sterility were achieved with GENESIS on experimental scale. Weather survey was made to identify suitable locations in Mexico and pilot seed production experiments were conducted at various locations. In the studies, the acceptable female sterility levels were observed along with low seed set in general, commercially unviable and variable results with locations. These studies concluded that heterosis is present at moderate to low levels. The good combining male and female backgrounds were identified. In addition, the heterotic pattern in CIMMYT lines was identified and the whole germplasm was divided into two distinct heterotic groups. After this, the work on CHA based hybrids or on hybrid development was discontinued at CIMMYT.

Indian Effort on Hybrid Wheat Development

In 1995, the Directorate of Wheat Research, Karnal initiated hybrid wheat development through CMS and CHA approach in network mode. Through CMS approach, cytoplasmic male sterile lines were developed by using *Triticum timopheevii, T. araraticum, Ae. caudata* and *Ae. speltoides* as source parents. No apparent adverse effects on morphology by *T. araraticum* and *Ae. speltoides* have been noticed. Two exotic genetic stocks registered as PWR 4099 and PWR 4101 indicated complete fertility restoration in *T. timopheevii* based CMS lines. Although there is no significant progress for heterosis for yield in totality, few hybrids showed heterosis for yield components, *viz.*, spikelet number, spike length and tillers/plant. Through CHA approach, chemicals were identified that induced male sterility when sprayed at 10–15 mm spike length 50–60 days after germination in most of the genotypes (Mahajan *et al.*, 1997). Over 1500 single cross hybrids developed through CHA were evaluated of which 6 hybrids were identified depicting standard heterosis over the best check and superior quality parameters. After thorough experimentation, 2 : 3 row ratio of male and female was found most suitable for high seed production. The profuse growth habit of 'oat' is used as space isolation to separate the various combinations of hybrid plots to avoid contamination from non-parental male genotypes. Supplementary pollination by rope pulling during the peak hours of flowering in the morning and afternoon was suggested for enhanced out-crossing. An "Improved DWR CHA spray system" was also fabricated for large scale hybrid seed production which has precise delivery of CHA with coverage rate of one acre per hour. Besides, a seed sowing drill was fabricated for simultaneous sowing of male and female parents for hybrid seed production.

Based on the results and experience in the Indian hybrid programme, it was concluded that out of the two pollination control systems (CHA and CMS), the cytoplasmic-genetic male sterility system, though cumbersome, provides useful and eco-friendly approach for hybrid seed production. The CHA approach indicated certain inherent problems beyond control and, therefore, not found feasible to produce hybrid seed. Besides, the CHA system was effective to attempt large number of hybrid combinations and preliminary testing of these hybrids on multilocation. The multilocation and multiyear data revealed that only few hybrids performed well. The parents of these superior hybrids might be used as basic material for CMS line development and restorer conversion. The *timopheevii* based CMS system was found to be most appropriate since the fertility restoration was highly satisfactory.

Future Strategies of Hybrid Wheat Research

For a successful hybrid breeding programme and commercially viable hybrid seed production, the following requirements must be met:

— The hybrids must satisfy the needs of the customer for all important traits. Simply to be "hybrid", or simply to exhibit "heterosis", is not enough.

— The price of hybrid seed must be low enough to enable the customer to make substantial profits from annually recurring investments in expensive hybrid seed. A rule of thumb is that the first time use of hybrid seed should enable the farmer to earn an extra profit equal to at least three times the additional cost of the seed.

— The price of hybrid seed must be high enough to enable the seed producer to make substantial profits from its investments in research, production, and sales.

To satisfy these primary criteria for success in the hybrid seed business, the organizations involved in hybrid development programme must integrate a host of variables such as:

- The pollinating system of the crop
- Options for manipulation of the pollinating system
- Supply and cost of labour for emasculation/other requirements for hybridization
- The yield of the crop in the farmer's field
- The commercial value of the crop per unit of land area
- The seeding rate of the crop
- The seed yield in the seed production field
- The extra yield to be expected from heterosis
- The implications of hybrid uniformity
- The most important traits to improve in the crop and their genetics
- The ease of demonstrating improvements in new hybrids, and
- Availability of inbred parents and other breeding materials in either public or private institutions.

Wheat hybrids can yield up to 30% more than their parents, but hybrids with heterosis at these levels usually are the product of crosses between different classes of wheat, such as a cross of hard red winter wheat with soft red winter wheat. Commercially useful wheat hybrids must be made within a class to maintain milling and baking quality. Crosses within a quality class typically have less heterosis, only about 5–15% more than their parents. The lower amount of heterosis may be because of relationship among members of a relatively closed gene pool. The experiments indicated that the heterosis level should be minimum 20% over the best check of the area for its commercial viability. As wheat is a self-pollinated crop, it has perfect florets, limited supplies of pollen, and a relatively brief period of stigma receptivity, hand emasculation is impractical for commercial production of hybrid seed, but cytoplasmic male sterility allows production of hybrid wheat seed on a field scale. Limited pollen production by the male lines as compared to maize, for example, means that the ratio of male rows to female rows must be relatively high, and seed yield per hectare is reduced correspondingly. If a hybrid has only a small yield advantage over the best pure line cultivars, the expected gain in the farmer's income from increased yield of the hybrid could be less than the cost of the hybrid seed, hence such a hybrid would be unacceptable despite its yield advantage. Breeders had to develop entirely new male lines, by inserting nuclear genes for fertility restoration into non-restorer genotypes. Restorer lines were usually made by introgressing dominant fertility restorer genes from widely divergent germplasm into elite wheat lines. Typically, the strongest and most useful restorer genes came from different species, sometimes weedy or wild species. This introduced problems of linkage to undesirable traits from the alien species. Breeders devoted much time and energy in backcrossing and selection for strong fertility restoration and, as a consequence, spent less time on breeding for increased yield and general performance. Also the public wheat breeding programmes provides very little input to hybrid wheat breeding in terms of investment. This led to under-investment in the development of germplasm and breeding methods (particularly for restorer males) in the important start-up period.

Interest in hybrid wheat is still present particularly in regions where wheat yields and commercial value of the crop are relatively high (Edwards, 1997). However, the hybrids are planted at very low seeding rates, thus keeping seed cost in line with the expected return. Research is in progress in several organisations on new ways to produce hybrid wheat seed, using new sterility systems, some of which are introduced into wheat via genetic transformation. The goal is to build systems that are reliable, easy to manipulate, which interfere as little as possible with the routine wheat breeding programs aimed at making improvements in yield and general performance. These attempts show that breeders still believe that hybrid wheat can succeed on large scale, and perhaps more importantly they show that there are numerous ways to produce the hybrids, and then to manage them for profit in the farmers' fields.

As the Directorate of Wheat Research is the nodal centre for wheat research and coordination in the country, it addressed the hybrid development programme through CMS system in network mode with cooperating centres in four major wheat zones, *i.e.* NWPZ, NEPZ, CZ and PZ. As most of the available CMS lines are in exotic background, there is need to diversify them in indigenous background for exploiting better adaptability of superior agronomic backgrounds. Similarly, the fertility restoration is an important issue which needs identification of fertility restorer lines and their diversification. The identification of heterotic groups for diversification programme is a pre-requisite for selection of diverse parents. As the level of heterosis is not commercially viable in most of the combinations, the future work needs for intra-pool improvement of male and female parental lines separately through pre-breeding approach so that distinct heterotic pools are obtained for realizing high heterosis levels. The search for open pollinating traits in diverse wheat germplasm lines and their exploitation and standardization of the hybrid seed production technology for economic seed production are the areas of future in the emphasized direction of development of a successful hybrid wheat technology. In nutshell, the following points may be considered as future strategy for hybrid wheat development programme.

- Identification of heterotic gene pools
- Creation of novel genetic variability for yield component traits from secondary and tertiary gene pools and its evaluation
- Improving fertility restoration by accumulation of Rf genes
- Use of variability available in the dwarfing genes for parental diversity
- Biotechnological tools for molecular interventions related to fertility restoration
- Search for heterosis in diverse gene pools
- Fixation of heterosis through apomixis
- Understanding of floral biology
- Promotion of active interaction between public and private organizations.

REFERENCES

Araki, H.1990. Studies on the practical use of hybrid wheat using cytoplasmic male sterility. *Bulletin of the Kyushu National Agricultural Experiment Station*, **26(2)**: 115–65.

Agarwala, S.C., P.N., Sharma, C., Chatterjee and C.P., Sharma. 1980. Copper deficiency induced changes in wheat anthers. *Proc. Indian Natl.Sci. Acad.Sect.* B2: 172–6.

Apltaverova, M., Z., Stehno and L., Dotlacil. 1981. Cytoplasmic male sterility in wheat. Vedeck'e Prace Vyzkumnho Ustavu Rostlinne Vyroby v Praze-Ruzyni, **21**: 29–52.

Athwal, D.S., P.S., Phull and J.L., Minocha. 1967. Genetic male sterility in wheat. *Euphytica,* **16**: 354–60.

Azouaou, Z. and A., Souvre. 1993. Effects of copper deficiency on pollen fertility and nucleic acids in durum wheat anther. *Sexual Plant Repro.*, **6**: 199–204.

Beri, S.M. and S.C., Anand. 1971. Factors affecting pollen-shedding capacity in wheat. *Euphytica,* **20**: 327–32.

Bingham, J. and E.G.H., Lupton. 1987. Production of new varieties: an integrated research approach to plant breeding. *In:* Wheat Breeding, Its Scientific Basis. ed., Lupton, E. G. H., Chapman and Hall, London and New York. pp. 487–538.

Bing-Hua, L. and D., Jing-Yang. 1986. A dominant gene for male sterility in wheat. *Plant Breeding* **97**: 204–9.

Borghi, B. and M., Perenzin. 1994. Diallel analysis to predict heterosis and combining ability for grain yield, yield components and bread making quality in bread wheat (*Triticum aestivum* L.). *Thero. Appl.Genet.,* **89(7–8)**: 975–81.

Bozzini, A. and G.T., Scarascia-Mugnozza. 1968. A factor for male sterility inherited as a Mendelian recessive. *Euphytica,* Supplement No. **1**: 83–86.

Briggle, L.W. 1963. Heterosis in wheat—A review. *Crop Sci.,* **3**: 407–12.

Briggle, L.W. 1970. A recessive gene for male sterility in hexaploid wheat. *Crop Sci.,* **10**: 693–6.

Chandra, S. and S.P., Bhatnagar. 1974. Reproductive biology of *Triticum*. II. Pollen germination, pollen tube growth, and its entry into the ovule. *Phytomorphology,* **24**: 211–7.

Chaudhary, H.K., A.S., Kapoor and D.P., Walia. 1993. Manifestation of heterosis in winter × spring wheat crosses, Short comm. in Symp. on Heterosis Breeding in Crop Plants, PAU, Ludhiana.

Chelak, V.R. 1989. Type of flowering and mode of pollination/fertilization in wheat (*Triticum aestivum* L.). *Botanicheskie Issledovaniya*, **4**: 11–30.

Chen, Q.F. 2003. Improving male fertility restoration of common wheat *Triticum timopheevii* cytoplasm. *Plant Breeding,* **122**: 401–4.

Chen, Q.F., Y.H., Zhou, Z.S., Peng and H.R., Jiang. 1998. Studies on the distribution of hybrid chlorosis Ch_1 gene and the T-type cytoplasm fertility restoring genes in Chinese endemic wheats. *Guihaiha,* **18**: 325–30.

Chhabra, A.K. and S.K., Sethi. 1991. Inheritance of cleistogamic flowering in durum wheat (*Triticum durum*). *Euphytica*, **55(2)**: 147–50.

Chhuneja, P. and J.L., Minocha. 1993. Induction of male sterility mutations in three cultivars of *Triticum aestivum* (L.). *In:* Heterosis Breeding in Crop Plants Theory and Application. Short communications: Symposium, Ludhiana, 23–24 February 1993 [*Edited* by Verma, M.M., Virk, D.S. and Chahal, G.S., Ludhiana, India; Crop Improvement Society of India, pp. 130–1.

Chowdhary, M. A., N., Mahmood and I., Khaliq. 1994. Pollen production studies in common bread wheat. *Rachis*, **11(1/2)**: 68–72.

Curtis, B.C. and D.R., Johnston. 1969. Hybrid wheat. *Sci. Amer.*, **220(5)**: 21–29.

D'Souza, V. L. 1970. Investigations concerning the suitability of wheat as pollen donor for cross-pollination by wind as compared to rye, *Triticum*, and *Secolo tricum*. (In German) *Z.Pflanzenzuecht*. **63**: 246–69.

De Vries, A.P.H. 1971. Flowering biology of wheat, particularly in view of hybrid seed production—a review. *Euphytica,* **20**: 152–70.

Driscoll, C.J. 1972. XYZ system of producing hybrid wheat. *Crop Sci.,* **12**: 516–7.

Driscoll, C.J. 1977. Registration of Cornerstone male-sterile wheat germplasm. *Crop Sci.,* **17**: 190.

Driscoll, C.J. 1979. Induction and use of the 'Cornerstone' male-sterility mutant in wheat. Proceedings of the Fifth International Wheat Genetics Symposium, India. 1978. pp. 499–502.

Driscoll, C.J. 1986. Nuclear male sterility systems in seed production of hybrid varieties. *CRC Critical Reviews in Plant Sciences* **3**: 227–56.

Du, H.L., S.S., Maan and J.J., Hammond. 1991. Genetic analysis of male-fertility restoration in wheat. III : Effects of aneuploidy. *Crop Science*, **31(2)**: 319–22.

DWR, 2007. Vision 2025. pp. 24.

East, E.M. and H.K., Hayes. 1912. Heterozygosis in evolution and in plant breeding. United States Department of Agriculture, Bureau of Plant Industries Bulletin **243**: 58.

Edwards, I. 1997. Personal communication. Pioneer Hi-Bred International, Inc., Johnston, IA, 50131 (USA). July 29, 1997.

Edwards, I.B. 2001. Hybrid wheat. *In*: A.P., Bonjean and W.J., Angus (*eds*), The World Wheat Book, a History of Wheat Breeding, Lavoisier. Inc., Paris 1017–45.

El-Kadi, D.A., L.H.A. Hindi, A.A., Gommaa and H.M., Ghazal. 1983. Analysis of male sterility in nine wheat hybrids with *T. timopheevii* cytoplasm. *Egyptian J. Genet. Cytol.*, **12(1)**: 1–15.

Engledow, F.L. and B.P., Pal. 1934. Investigations on yield in cereals. VIII. Hybrid vigour in wheat. *Journal of Agricultural Science,* **24**: 390–409.

Evans, L.T., J., Bingham and M.A., Roskams. 1972. The pattern of grain set within ears of wheat. *Australian Journal of Biological Sciences,* **25**: 1–8.

Fedin, M.A. 1976. Heterosis in wheat and its hereditary factors. *In:* Heterosis in Plant Breeding, Proceedings of the Seventh Congress of EUCARPIA. *eds*., Janossy, A. and E.G.H., Lupton, Elsevier, pp. 291–302.

Fossati, A. and M., Ingold. 1970. A male sterile mutant in *Triticum aestivum. Wheat Information Service,* **30**: 8–10.

Freeman, G.F. 1919. Heredity of quantitative characters in wheat. *Genetics*, **4**: 1–9.

Gorin, A.P. 1968. Results of studies on the biology of flowering and pollination in field crops. *Dokl. Sel.-Khoz. Akad. Timiryazeva*, **139**: 175–181 (*Pl. Breed. Abst.* 39, 1969; 4289).

Gotsova, D.K., I., Panayatov and K., Gotsov. 1987. Male fertility restoration against various alien cytoplasms. II. Genetical analysis of R-17127. *Wheat Inf. Ser.*, **64**: 1–5.

Griffin, W.B. 1987. *N.Z.J. Agric Res.*, **30**: 287–90.

Halloran, G.M. 1975a. Heterosis in wheat. *Zeitschriftjar Pjlanzenzuchtung,* **74**: 18–27.

Halloran, G.M. 1975b. Genetic analysis of yield in wheat. *Zeitschrift jar Pjlanzenzuchtung,* **74**: 298–321.

Hermsen, J.G. Th. 1965. Towards a more efficient utilization of genic male sterility in breeding hybrid barley and wheat. *Euphytica,* **14**: 221–4.

Heyne, E.G. and G.S., Smith. 1967. *In:* "Wheat and wheat Improvement." (*eds*., K.S., Quisenberry and L.P., Reitz). *Amer. Soc. Agron.*, Madison, Wisconsin, pp. 269–306.

Hoshikawa, K. 1960. Studies on the reopened floret in wheat. *Proceedings of Crop Science Society of Japan,* **29**: 103–6.

Hsu, P. and P.D., Walton. 1970. The inheritance of morphological and agronomic characters in spring wheat. *Euphytica,* **19**: 54–60.

Huang T.C. (ed). 1990. Studies on Hybrid Wheat-Progress, Problems and Prospects. Publishing House of Beijing Agricultural University, Beijing, China.

Hucl, P. 1996. Out crossing rates for 10 Canadian spring wheat cultivars. *Canadian J. Plant Sci.,* **76(3)**: 423–7.

Ikeguchi, S., A., Hasegawa, Y., Oyamada, K., Toriyama and K., Tsunewaki. 1994. Basic studies on hybrid wheat breeding utilizing *Aegilops kotschyi* cytoplasm. *In:* Proceedings of the Japan-Russia Workshop Sapporo on: Low Temperature Physiology and

Breeding of Northern Crops, Sapporo, Japan, January 25–28, 1994. Hokkaido National Agricultural Experiment Station, 77–83.

Il'ina, L.B. 1984. Inheritance of genic male sterility in bread wheat. *In:* Genet. Selekts. Issled. na Urale. Irf. Materialy Sverdlovsk, USSR, 48–50.

Il'ina, L.B. 1985. Investigating CMS in the process of producing sterile analogues of spring bread wheat. Ufa, USSR Rost I Produktivnost Rast. Dep., 2136–**85**: 15–32.

Jan, C. C. 1974. Genetic Male Sterility in Wheat (*Triticum aestivum* L): Expression, Stability, Inheritance and Practical Use. Ph.D thesis, University of California, Davis.

Jan, C.C. and C.O., Qualset. 1977. Genetic male sterility in wheat *(Triticum aestivum* L.): reproductive characteristics and possible use in hybrid wheat breeding. *Hilgardia,* **45**: 153–61.

Jewell, A.W., B.G., Murray and Alloway. 1988. Light and electron microscopic studies on pollen development in barley grown under copper sufficient and deficient conditions. *Plant Cell Environ.*, **11**: 273–81.

Karim, M.A. and M.P. Singh. 1984. Studies on fertility restoration in male sterile wheats derived from *Aegilops comosa* cytoplasm. *Wheat Inf. Ser.,* **58**: 9–11.

Kaul, M.L.H. 1988. Male sterility in higher plants. *Monographs on Theoretical and Applied Genetics* 10. Springer-Verlag, Berlin.

Kihara, H. 1951. Substitution of nucleus and its effects on genome manifestations. *Cytologia,* **16**: 177–93.

Kihara, H. and K., Tsunewaki. 1964. Some fundamental problems underlying the program for hybrid wheat breeding. *Seiken Ziho.,* **16**: 1–14.

Kirby, E.J.M. and M., Appleyard. 1981. *Cereal Development Guide.* Second edition. NAC Cereal Unit, Stoneleigh.

Komaki, M.K. and K., Tsunewaki. 1981. Genetical studies on the difference of anther length among common wheat cultivars *Euphytica*, **30**: 45–53.

Kucera, L. 1982. Monosomic analysis of fertility in common wheat. Prof. Marchal. *Euphytica,* **31**: 895–900.

Kumar, Shreekant and D.K. Ganguli. 1993. Heterosis and inbreeding depression in bread wheat. *In:* Heterosis Breeding in Crop Plants. Theory and Application. Short communication. Symposium, Ludhiana, 23–24 Feb., 1993. [*eds.*, Verma, M.M., Virk, D.S. and Chahal, G.S.] *Crop Improvement Society of India*, 62–63 Ludhiana, India.

Larik, A.S., A.R., Mahar and H.M.I., Hafiz. 1995. Heterosis and combining ability estimates in diallel crosses of six cultivars of spring wheat. *Wheat Inf. Ser.*, **80**: 12–19.

Leighty, C. E. and J. W. Taylor. 1927. Studies in natural hybridization of wheat. *J. Amer. Soc. Agron.*, **10**: 865–87.

Leighty, C. E. and W. J. Sando. 1924. The blooming of wheat flowers. *J. Agric. Res.*, 27: 231–44.

Lelley, J. 1966. Observation on the biology of fertilization with regard to seed production in hybrid wheat. (in German) *Der Zuchter*, **36**: 314–7.

Liable, C.A. 1974. Chemical methods for pollen control. *Proc. 29th Ann. Corn and Sorghum Res. Conf. Pub.*, **29**: 174–82.

Liu, B.H. and L., Yang. 1994. Dwarfing sterile wheat and its use in breeding for dwarfness. *Scientia Agri. Sinica*, **27(5)**: 17–21.

Liu, Bing Hua, Shan Hong Wang and Li Yang. 1996. The discovery and identification of new material of nuclear male sterility in common wheat. *Hereditas (Beijing),* **18**(6): 9–11.

Luo, H.B., J.M., He, J.T., Dai, X.G., Liu and Y.C., Yang. 1998. Studies on the characteristics of seed production of two ecological male sterile lines in wheat. J. *Hunan Agril Univ.*, **24(2)**: 83–89.

Lupton, E.G.H. 1961. Studies in the breeding of self-pollinating cereals. 3. Further studies in cross prediction. *Euphytica,* **10**: 209–24.

Lupton, E.G.H., J., Bingham, J.A., Blackman, W.J., Angus, P.J., Jackson, R.H., Oliver, A., Kirby and M., Taylor. 1980. Nuclear factor for male sterility. *Annual Report Plant Breeding Institute, Cambridge* 1979: 21.

Ma, Z.Q., Y.H., Zhao and D.J., Liu. 1991. Incorporation of restoring gene of *Aegilops umbellulata* into wheat. *Genome,* **34(5)**: 727–32.

Macer, R.C.E. 1972. The resistance of cereals to yellow rust and its exploitation by plant breeding. *Proceedings of the Royal Society, London B,* **181**: 281–301.

Mahajan, Vinay, Kuldeep Singh, and R.G., Kelkar. 1997. On the possibility of using CHA route in developing wheat hybrids. *In:* Abstract of Int. Group Meeting on Wheat Research on Needs Beyond 2000 AD. Karnal, pp. 41–42.

Mann, S.S. 1973. Cytoplasmic male sterility and male fertility restoration system in wheat. Eucarpia Meeting (Cereal Section). Institute of Plant Breeding, Bari, Italy, **54**: 14–18.

Mann, S.S. 1992. Genetic analysis for male fertility restoration in wheat. V. anomalous results of monosomics analysis. *Crop Sci.,* **32**: 28–35.

Martin, T.J. 1990. Outcrossing in twelve hard red winter wheat cultivars. *Crop Sci.,* **30**: 59–62.

McDaniel, R.G. and I.V., Sarkissian. 1966. Heterosis: complementation by mitochondria. *Science,* **152**: 1640–2.

McIntosh, R.A. and J.E., Cusick. 1987. Linkage map of hexaploid wheat. *In:* E.G. Heyne (*ed.*) *Wheat and Wheat Improvement-Agronomy Monograph No. 13*. 2nd edn., 289–97. Wisconsin, USA, ASA-CSSASSSA.

Miller, J.F. and K.A., Lucken. 1976. Hybrid wheat seed production methods for North Dakota. *Crop Sci.,* **16**: 217–21.

Molnar Lang, M.B., Barnabas and E., Rajki 1980. Changes in the shape, volume weight and the tissue structure of the pistil in the flowers of male sterile wheats during flowering. *Cereals Res. Comun.,* **8**: 371–9.

Murai, K. 1997a. Effects of *Aegilops crassa* cytoplasm on the agronomic characters in photoperiod-sensitive CMS wheat lines and F_1 hybrids. *Breeding Science,* **47**: 321–6.

Murai, K. 1997b. Genetic analysis of fertility restoration against photoperiod-sensitive cytoplasmic male sterility in *Triticum aestivum* cv. 'Norin 61'. *Plant Breeding,* **116**: 592–4.

Murai, K. 2002. Comparison of two fertility restoration systems against photoperiod-sensitive cytoplasmic male sterility in wheat. *Plant Breeding,* **121**: 363–5.

Murai, K. and K., Tsunewaki. 1994. Genetic analysis on the fertility restoration by *Triticum aestivum* cv. Chinese Spring against photoperiod-sensitive cytoplasmic male sterility. *Japanese Journal of Genetics,* **69**: 195–202.

Murai, K. and K., Tsunewaki. 1995. Photoperiod-sensitive Cytoplasmic male sterility induced in Japanese wheat cultivars by transferring *Aegilops crassa* cytoplasm *Breeding Science*. **45**: 199–203.

Murai, K., Y., Ogihara and K., Tsunewaki. 1995. An EMS-induced wheat mutant restoring fertility against photoperiod-sensitive cytoplasmic male sterility. *Plant Breeding,* **114(3)**: 205–9.

Nettevich, E.D. and A.A., Naumov. 1970. The genetic characteristics of fertility restoration in wheat forms with cytoplasmic male sterility. Nauch. tr. N IIs. kh. tsentr. r-nov nechernozemn-zony, **25(I)**: 77–85.

Newton, K.J. 1988. Plant mitochondrial genomes: organisation, expression and variation. *Annual Review of Plant Physiology and Plant Molecular Biology,* **39**: 503–32.

OECD. 1991. OECD scheme for the varietal certification of cereal seed moving in international trade. Organization for Economic Co-operation and Development, Paris.

Oehler, E. and M., Ingold. 1966. New cases of male sterility and new restorer source in *T.aestivum*. *Wheat Inf. Serv. (Kyoto)*, **22**: 1–3.

Panayotov, I., D. K. Gotsova and K. Gotsov. 1986. Male fertility restoration against alien cytoplasm. I. Comparison between the restoration abilities of three groups of lines. *Wheat Inf. Ser.*, **63**: 7–10.

Percival, J. 1921. The Wheat Plant—A Monograph. Duchworth & Co. London, 463.

Pickett, A. A. 1993. Hybrid wheat: results and problems. *Adv. Plant Breed., Suppl. J. Plant Breeding*, **15**: 1–259.

Picket, A.A, and N.W. Galway. 1997. A further evaluation of hybrid wheat. *Plant Cultivars and Seeds*, **10**: 15–32.

Pinthus, M.L. and E., Millet. 1978. Interactions among number of spikelets, number of grains and grain weight in the spikes of wheat *(Triticum aestivum* L.). *Annals of Botany*, **42**: 839–48.

Pugsley, A.T. and R.N., Oram. 1959. Genic male sterility in wheat. *Australian Plant Breeding and Genetics Newsletter*, 14.

Qian, C.M., A., Xu and G.H., Uiang. 1986. Effects of low temperatures and genotypes on pollen development in wheat. *Crop Science*, **26**: 43–46.

Ralko, V.P. 1982. Effect of the maternal genotype on degree of pollen ferility restoration in wheat hybrids bred using male sterility. *Vestsi AN BSSR*, Ser Biyaln. No. 3, 33–38.

Rawson, H.M. 1970. Spikelet number, its control and relation to yield per ear in wheat. *Australian Journal of Biological Sciences*, **23**: 1–15.

Rerkasem, B. and S., Jamjod. 1997. Boron deficiency induced ear sterility in wheat (*T. aestivum* L.) and its implication for plant breeding. *Euphytica*, **96**: 257–62.

Rodriguez, R, M.A., Quinones, N.E., Borlaug and I., Narvaez. 1967. Hybrid wheats: their development and food potential. *Research Bulletin* 3. International Maize and Wheat Imprpvement Center, Londres, Mexico.

Sage, G.C.M. 1976. Nucleo-cytoplasmic relationships in wheat. *Advances in Agronomy*, **28**: 267–300.

Schmidt, J.W., V.A., Johnson and S. S., Maan. 1962. Hybrid wheat. *Nebraska Experimental Station Quarterly*, **9**: 9.

Shaowen, Y. and R., Shan. 1980. Studies on the VE-type male sterility of wheat. *Acta Genetica Sinica*, **7**: 26–35.

Singh, S., R., Dhari and A.K., Joshi. 1997. Expression of heterosis for yield and yield traits in Indian wheat crosses under drill sown condition, CIMMYT 1997. Book of Abstracts. The Genetics and Exploitation of Heterosis in Crops; An International Symposium, Mexico, D.F., Mexico, pp. 336–7.

Singh, S.K. 2003. Cluster analysis for heterosis in wheat [*Triticum aestivum* (L.) em Thell.]. *Indian J. Genet.*, **63(3)**: 249–50.

Singh, S.K. 2005. Character association and path coefficient analysis for yield and floral characters in wheat (*Triticum aestivum* L. em Thell). *Crop Improv.*, **32(2)**: 124–9.

Singh, S.K. 2006. Evaluation of spring wheat [*Triticum aestivum* (L.) em Thell] germplasm for various floral characteristics. *SAARC J. Agri.*, **4**: 167–77.

Singh, S.K. 2007. Germplasm developed for high 1000-grain weight and other yield component traits. DWR News. Jan-June, 2007. pp 7.

Singh, S.K. and A.K., Joshi. 2003. Variability and character association for various floral characters in wheat [*Triticum aestivum (L.)* em Thell.]. *Indian J. Genet.*, **63(2)**: 153–4.

Singh, S. K., A.K., Joshi and B., Arun. 2007. Comparative evaluation of exotic and adapted germplasm of spring wheat for floral characteristics in the Indo-Ganetic Plains of India. *Plant Breeding*, **126**: 559–64.

Singh, S.K. and R.M., Singh. 2001. Variability and character association among floral traits and yield in bread wheat [*Triticum aestivum* (L.) em Thell]. *Indian J. Plant Genet. Resour.*, **14(2)**: 199–201.

Singh, S.K. and R.M., Singh. 2004. Floral biology and character association studies in wheat. *Farm Sci. J.*, **13(1)**: 21–23.

Singh, S.K. and R.M., Singh. 2006. Determination of optimum row ratio for hybrid seed production in wheat. *Annual Wheat Newsletter,* **52**: 56–57.

Singh, S.K., R.M., Singh, A.K., Joshi and R., Dhari. 2006. Effect of etherel on seed setting, outcrossing, pollen sterility and yield traits in wheat (*Triticum aestivum* (L.) Em Thell]. *Annual Wheat Newsletter*, **52**: 51–56.

Snape, J.W. 1982. Predicting the frequencies of transgressive segregants for yield and yield components in wheat. *Theoretical and Applied Genetics,* **62**: 127–34.

Sneep, J., B.R., Murty and H.F., Utz. 1979. Current breeding methods. *In*: *Plant Breeding Perspectives. Eds*: Sneep, Hendriksen, A.T. and Holbek, O. Pub., Centre for Agricultural Publishing and Documentation, Wageningen, pp. 104–233.

Stehno, Z. and M., Apltaverova. 1980. Pollen sterility induced in wheat (*Triticum aetivum* L.) by the cytoplasm of *Aegilops sp. Sci. Agri. Bohenoslovaca*, **12(4)**: 245–53.

Stehno, Z. and M., Apltaverova. 1981. The effect of the cytoplasm of alloplasmic lines of wheat (*Triticum aestivum*). *Sci. Agri. Bohemoslovaca*, **13(4)**: 281–9.

Stroike, J.E. 1987. Technical and economic aspects of hybrid wheat seed production. *In*: *Hybrid Seed Production of Selected Cereal, Oil and Vegetable Crops, FAO Plant Protection and Production Paper* 82. Food and Agriculture Organisation of the United Nations, Rome, pp. 177–85.

Sylvester-Bradley, R. and R.K., Scott. 1990. Physiology in the production and improvement of cereals. *HGCA Research Review No.* 18. Home-Grown Cereals Authority, London.

Tsujimoto, J. and K., Tsunewaki. 1984. Chromosome location of fertility restoring gene of a common wheat Chinese Spring for the *Aegilops mutica* cytoplasm. *Wheat Inf. Ser.*, **58**: 4–8.

Tsunewaki, K., T.R., Endo, M., Kobayashi, Y., Mukai and I., Panayotov. 1980. *Genetic Diversity of the Cytoplasm in Triticum and Aegilops*. Japan Society for the Promotion of Science, Tokyo.

Tsunewaki, K. 1982. Monosomic analysis on the fertility restoration by *T. aestivum* cv. Chinese Spring against *Aegilops ovata* cytoplasm. *Japanese J. Genet.*, **57(5)**: 513–25.

USDA, 2007. Grain: World markets and trade. Dec, 2007. pp. 7.

Ushiyama, T., K., Toriyama, K., Tsunewaki, S., Nonaka, T., Shimada and S., Nonaka. 1996. Seed production ability of male sterile lines of common wheat induced by the interaction between an SV type cytoplasm and a 1BL 1RS chromosome. *Breed. Sci.*, **46(3)**: 303–6.

Virmani, S.S. and I.B. Edwards. 1983. Current status and future prospects for breeding hybrid rice and wheat. *Advances in Agronomy,* **36**: 145–214.

Wang, Yan, Shu Ren Liu, Yui Jie Zhai and Yanling Diao. 1997. Studies on testing and utilization of heterosis in spring wheat hybrids. *Acta Agri. Boreali-Sinica*, **12(1)**: 17–21.

Whingwiri, E.E. and W.R., Stern. 1982. Floret survival in wheat: significance of the time of floret initiation relative to terminal spikelet formation. *Journal of Agricultural Science,* **98**: 257–68.

Whitehouse, R.N.H., J.B., Thompson and M.A.M., Do Valle Ribeiro. 1958. Studies on the breeding of self-pollinating cereals. 2. The use of a diallel cross analysis in yield prediction. *Euphytica,* **7**: 147–69.

Wilson, J.A. 1968. Problems in hybrid wheat breading. *Euphytica*,**14**, Suppl., **1**: 13–33.

Wilson, J.A. 1972. Hybrid Wheat Breeding. Rice Breeding, International Rice Research Station, Manila, Philippines, pp. 593–602.

Wilson, J.A. 1984. Hybrid wheat breeding and commercial seed development. *Plant Breeding Reviews,* **2**: 303–19.

Wilson, J.A. and W.M., Ross. 1961. Cross breeding in wheat, *Triticum aestivum.* L Frequency of the pollen-restoring character in hybrid wheat having *Aegilops ovata* cytoplasm. *Crop Science,* **1**: 191–3.

Wilson J.A. and W. M., Ross. 1962. Male sterility interaction of the *Triticum timopheevi* cytoplasm. *Wheat Inf. Serv. (Kyoto)* University, **14**: 29–30.

Wilson, J.A. and W.M., Ross (1962a). Cross breeding in wheat, *Triticum aestivum* L. II. Hybrid seed set on a cytoplasmic male-sterile winter wheat composite subjected to cross-pollination. *Crop Science,* **2**: 415–7.

Wilson, J.A. 1968b. Hybrid wheat developments with *Triticum timopheevi* Zhuk. derivatives. *Proceedings of the Third International Wheat Genetics Symposium, Canberra,* 1968. pp. 423–9.

Wricke, G. 1989. Genetic mechanisms for hybrid seed production. *Vortriige fur Jllanzenzuchtung,* **16**: 369–78.

Wrobel, A. 1989. Genetic investigation of fertility restoration in male sterile forms of wheat. *Hodowla Roslin Aklimatyzacja i Nasiennictwo,* **29(3–4)**: 13–26.

Yu, Z.W., D.A., Van Sandford and D.B., Egli. 1988. The effect of population density on floret initiation, development and abortion in winter wheat. *Annals of Botany,* **62**: 295–302.

Zehr, B.E., V.P., Ratnalikar, L.M.M. Reddy and L.V., Pandey. 1997. Strategies for utilizing heterosis in wheat, rice and oilseed Brassica in India (Abstr B28) *In*: *The Genetics and Exploitation of Heterosis in Crops*, 17–22 Aug. 1997, Mexico City, Mexico, pp. 232–3.

Zeven, A.C. 1967. Transfer and 'inactivation of male sterility and sources of restorer genes in wheat'. *Euphytica,* **16**: 183–9.

Zhang, G.S. and T. Z., Yang. 1989. A preliminary study on male-sterile lines of wheat with *Aegilops ventricosa*, *A. kotschyi* and *A. variabilis* cytoplasms. *Acta Agron. Sinica,* **15(1)**: 1–10.

Zhou, M.L., Q.Y., Tang and J.M., He. 1997. Effect of photoperiod on fertility of ecotype male sterile (EMS) wheat. *In:* R.L., Lu, X.B., Cao, F.M., Liao and Y.Y., Xin (*eds*). Proceedings of the International Symposium on Two-Line System of Heterosis Breeding in Crops, 6–8 September 1997. Changsha, China, CNHRR&DC, Hunan pp. 107–13.

10 Promotion of Partnerships for Commercialization of Hybrid Technology

S.K. RAO*

ABSTRACT

Hybrid technology offers considerable opportunity for increasing productivity of major cereals, especially rice, maize, sorghum and pearl millet in India. Significant development in the country has demonstrated the potential of hybrids over high yielding varieties even under stress environments. Unfortunately, benefits of this technology have not reached the farmers on an extensive scale, especially for want of strong promotional activities. There is a need to promote hybrid technology to meet future challenges and to meet increasing demand for food grains and feeds. Hence, timely action to promote suitable hybrids would go a long way in increasing crop productivity. Hybrid technology is obviously an appropriate technology for sustainable agriculture, and thus deserves to be promoted on a large scale to ensure food security.

Key words: Heterosis, linkage, limiting factors male sterility system, trade liberalization

Seed is the most crucial, critical and vital input for attaining sustained growth in agricultural production. Hybrid leads to vertical increase in agricultural production because of much higher yield potential compared to high yielding open-pollinated varieties. However, it requires a lot more attention to harness the immense potential in different crops.

Heterosis, the genetic principle on which hybrids are based, is now a recognized means to boost crop productivity. Hybrids with economic levels of commercial heterosis for yield and consumer quality have been accepted widely. Ensured wide acceptance among farmers is equally important in hybrids apart from creating hybrid vigour through effective plant breeding research. It is important to recognize that, research and development are two phases of productivity improvement efforts that need to be analysed simultaneously.

* Director, Directorate of Farms, JNKVV, Jabalpur - 482 004.

Private sector has an imperative role in agricultural productivity improvements, earlier through supply of other agricultural inputs, but now, through the supply of core input-the seed.

Development

The success of hybrid maize provided necessary impetus to the breeders of other crops, including sorghum and pearlmillet. Accordingly, the All India Coordinated Projects were launched that resulted in the development and release of number of hybrids for general cultivation in these two crops. Subsequently in 1968, Dr. C.G. Patel at Cotton Research Station, Surat demonstrated the possibility of commercial exploitation of hybrid vigour in self-pollinated crops like cotton by hand emasculation.

The introduction of cytoplasmic genetic male-sterility (CGMS) system provides a major breakthrough in hybrid seed production in many self-pollinated crops, *viz.*, sorghum (1964), pearlmillet (1965) and rice among cereals; cotton (1990) among fibre crops; sunflower (1980) and castor (1973) in oilseeds, pigeonpea (2006) among pulses and fodder sorghum (1981).

The availability of hybrid seed in the country from public sector organizations is very limited. In spite of distinct advantages of maize hybrids, the availability of hybrid seeds and area covered under it, in different states is unsatisfactory. This may be due to emphasis given on promoting pure line varieties rather than hybrids during the past two decades. Now it has been realized that promotion of hybrid seed technology is the only answer, as OPVs would not bring the required breakthrough in productivity. In maize, non-availability of high yielding single cross hybrids with 90 days or less maturity period is significant limiting factor whereas, the utilization of hybrids seed of cotton and sunflower is much less compared to other crops. Thus, steps need to be taken to encourage hybrid seed production by the public sector seed corporations.

Scope

Public sector seed organizations in Gujarat, Andhra Pradesh and Karnataka are producing hybrid seeds of pearlmillet, maize, cotton, castor and sunflower in comparatively larger quantities. Other State Seed Corporations have to augment hybrid seed production programme to increase productivity. At the same time, it would be necessary for the public research institutions to develop suitable hybrids required for varied agro-climate conditions prevailing in different states.

In spite of appreciable quantity of hybrid seeds production in Andhra Pradesh, Gujarat, Karnataka and Maharashtra the production remained static during the last 3–4 years. This petite hybrid share is not considered significant therefore seed corporations will have to go a long way in increasing the hybrid share and their profitability.

Factors

Limiting factors for popularization of hybrid seed by public organizations:

1. Establishment of superiority of hybrids over high yielding varieties required effective extension services by State Departments of Agriculture as the farmers are misinformed about the benefit of hybrid seeds. Hybrid seeds extension programme will be of immense help in growing consciousness and generating hybrid seed demand.
2. Production cost of hybrid seed may be subsidized, as its production is expensive.
3. A prerequisite for success of hybrid seed, especially in self-pollinated crops, is that the commercial yield increase should not only offset the additional seed cost but should be able to bring higher profit to the farmers.
4. Industrial use of the commercial produce has to be ensured to maintain the market price at remunerative level. Continued government and industrial assurance along with infrastructural support to accept higher production would be necessary to achieve it.
5. Seed replacement rate of hybrids in north India may be enhanced by introducing short duration maize (85–90 days) and sunflower (75–80 days) hybrids, as long duration hybrids do not fit in the prevailing cropping patterns.
6. In seed industry, demand assessment of hybrid seed is important for fixing unambiguous responsibilities and commitments.
7. To ensure continuous supply of parental materials, public sector organizations need to set up their own R&D units for assured multiplication of adequate quantities of quality seed of the parental lines.
8. The public sector organizations should collaborate at global level to lucratively explore superior hi-tech materials.

It is unfortunate that little and untimely effort has gone into the popularization of public sector hybrids developed through a reasonably well-planned infrastructure of institutions and research stations spread all over the country and released through a worldwide acknowledged system of multilocation coordinated testing. Supply of seed to the farmer remained a constant problem ever since organized seed production was initiated in the country in the mid-sixties. Hybrids/inbred varieties released from the public system still occupy the dominant share in India's agricultural production but it is taking an awfully long period of 7–10 years before the newly released materials become available to a large number of farmers. It will not be surprising that many of the sick seed corporations may have to go out of business in future unless they cope up with the changes in scenario of seed trade.

Benefits

The larger gap between average crop productivity at the state or national level and the realized potential of new material can be reduced only through popularization of public-bred materials. In this endeavour, the Government of India may like to consider national hybrid seed planning by developing a system where in the very first year of the release of hybrid the seed is supplied in small packets up to village and panchayat level to bring awareness in time.

The public sector agriculture provides education, research and technologies through the national agriculture research systems. This system has generated hybrid technologies in various crops like rice, maize, sorghum, pearl millet, sunflower, pigeonpea, castor etc. Due to weak hybrid seed development plans at the national and state levels, seed developmental agencies could not produce enough quantity of seed to reach larger areas. The state seed extension support systems could not meet the demands of the farming communities. Hence the future of agriculture rests with hybrid seed along with hybrid technologies. A dramatic change in the socio-economic condition of the farmers has been noticed with popularization and commercialization of hybrid technologies.

Possible Threat

Private sector focuses on commercial product development, generating money through hybrid seed sales with high level of accountability to the customers as well as stakeholders. This sector conducts focused applied research that may provide returns on investment through IPR and help further promotion of investments. Now it is a necessity to conduct large scale research to compete effectively by developing hybrids suitable for wide range of environments and customers by utilizing novel genetic and technological resources.

Meanwhile, the industry has undergone a wide range of transformations, the role of private seed companies getting importance and acquiring more market share in hybrid seeds. Establishing joint ventures of progressive seed companies with huge investments in biotechnology and change in the regulatory framework will bring major changes in the commercialization of hybrid technologies.

Several areas of cooperation have been identified, starting from sharing of genetic resources to seed extension services, training, etc in the popularization of hybrid technologies on the basis of mutual interest, trust and confidence between the partners.

The public-private partnerships will always aim at benefit to the consumers as well as environment. ICRISAT is one of the examples where research consortium concerts with private sector. The initiative of ITC in the form of E-*choupal* is another excellent example of partnership. For the year 2008–09 Government of Madhya Pradesh has given the responsibility of promoting extension activities to private organizations.

Institutes like IARI/ UAS Dharwad, JNKVV, GBPUA&T, KKV, TNAU, NDAUT and several other Agricultural Universities have developed partnerships with private sector in the promotion of hybrid technologies for commercialization. The best example is the rice hybrid PRH 10 of IARI that has successfully reached the farmers in far-off areas through public-private partnership mode. Similarly rice hybrid Shayadri of KKV Dapoli has seen the light of the day in the large rice growing areas through Syngenta. All the State Agriculture Universities and ICAR Institutes provided training for management of Bt crop and monitoring with a strong support for conducting agronomical and hybrid evaluation through various contract research programmes. After development of products, the commercialization aspects must be left to the private agencies. The private seed companies have been primarily limited to the hybrids with better agronomic, pest, insect, and disease resistance traits. Seed companies are increasing their research budgets and activities for the development of hybrids suitable for targeted environments. During the process of hybrid development several important considerations such as technological options in terms of land, capital, labor and knowledge have been made to manage various technologies with the change in the farming systems. Assessment to achieve the market demand and quality specification of the produce is a big task.

Trade liberalization has to be considered in the product selling under partnership mode. In the public-private partnership, the private agency is always interested in retaining management responsibility so that the partnership can be transferred to the more challenging task. In the partnership mode one has to look at the product development and delivery cycle to meet the consumers needs which includes research and development, production, processing, marketing, distribution and customer services. Under IPR regime there are multiple damages for consumers through increasing market concentration, high cost of seed for farmers, high prices for public as well as genetic uniformity due to non-sharing of genetic resources may hinder the crop improvement programme.

For successful promotion of partnerships, each partner should bring value for both the partners to the table with the common objectives with complimentary role for effective delivery of the produce to the consumers.

The driving factors of collaboration may increase innovation in agriculture biotechnology, globalization and trade liberalizations. However, major emphasis will be on product differentiation to be marketed with the global competitiveness.

The public sector organization must train plant breeders with strong background of marker-assisted breeding. Public sector may also conduct contractual research for private sector and provide genetic resources on the basis of MTA/MOU to encourage the promotion of successful partnerships.

Success of such partnership can easily be realized through specified products, farmer-driven programme and appropriate policy environments includes breeder to operate, free access to germplasm, segregation population, foundation seed and ability to fix a price based on value addition, market, access for produce, enabled IP environment which includes PPV&FR and patents.

CONCLUSION

In the promotion and success of partnerships, a clear understanding of rules and regulations based on multiple laws, policies, harmonization of law, policies, etc. are essential.

State governments as well as SAUs have to play a vital role to create an environment with the change of mind set from regulator to facilitator, less competition, more skill development and complementation to the business promotional activities. In the legal framework, partnerships have to make through single window in spirit and practice.

The public sector agricultural research must adjust to meet the challenges of privatization and attention must shift to specific opportunities for interaction with the private sector. Similarly, public plant breeding programmes with liberalized seed regulations must be favourable for partnership promotions to benefit the society.

Society will be best served by non-monopolistic mode of both private and public sector plant breeding that has both competitive and collaborative aspects.

N.W. Simmonds

Hence it could be concluded that hybrid technology offers considerable opportunity for increasing productivity of major cereals, especially rice, maize, sorghum and pearl millet in India. Significant development in the country has demonstrated the potential of hybrids over high yielding varieties even under stress environments. Unfortunately, benefits of this technology have not reached the farmers on an extensive scale, especially for want of strong promotional activities. There is a need to promote hybrid technology to meet future challenges and to meet increasing demand for food grains and feeds. Hence, timely action to promote suitable hybrids would go a long way in increasing crop productivity. Hybrid technology is obviously an appropriate technology for sustainable agriculture, and thus deserves to be promoted on a large scale to ensure food security.

11

Agribusiness in India: Opportunities and Prospects

P. Subbian*

ABSTRACT

Due to growth in population, abundance of productive and inexpensive land, problems created in farm labor market, squeeze on land holdings, and generation of new farm technologies and inputs prompted farmers to increasingly indulge in buying labor saving, and yield-increasing farm inputs from market. The marketed and marketable surplus on farm started growing, turning agriculture from subsistence to commercial. Business started building up in and around agriculture that results in the birth of agribusiness. The process starts at the producer level and reaches the final consumers through vertical integration. In this process, right decisions at the right time relating to production and post harvest management and managerial issues such as professional assistance in production, finance, marketing and government policies are indispensable.

Key words: Agribusiness, crop diversification, export promotion, information technology

Modern agriculture is the result of a variety of economic and technological forces. Since the mid-1960's, the pace of change has quickened. New technologies and industries have emerged. Traditional farming operations have grown larger and more specialized in most countries-developed and developing. Agriculture had been (in few developed) and is (in most of the developing countries) a major industry across the globe.

Growth in population, abundance of productive and inexpensive land, problems created in farm labor market, squeeze on land holdings, and generation of new farm technologies and inputs prompted farmers to increasingly indulge in buying labor saving, and yield-increasing farm inputs from market. Even some members of farm families started withdrawing from farm and moving to non-farm jobs. The marketed and

* Director, Directorate of Agribusiness Development, Tamil Nadu Agricultural University, Coimbatore - 641 003.

marketable surplus on farm started growing, turning agriculture from subsistence to commercial. Businesses started building up in and around agriculture. This resulted in the real birth and growth of agribusiness.

AGRIBUSINESS

Agribusiness encompasses a whole lot of activities of agricultural sector like integration of production, processing and marketing. The process starts at the producer level and reaches the final consumers through vertical integration. In this process, decision relating to production and post harvest management where storage, transportation, packaging and branding assume greater significance. Also taking right decisions at the right time, managerial issues such as professional assistance in production, finance, marketing and government policies are indispensable.

Agribusiness refers to the various businesses involved in food production, including farming, seed supply, agrichemicals, farm machinery, wholesale and distribution, processing, marketing, and retail sales.

Organizations Involved in Agribusiness

- Small Farmers Agribusiness Consortium (SFAC)
- Agricultural and Processed Food Products Export Development Authority (APEDA)
- National Horticultural Board (NHB)
- Commodity Boards: (a) Spice Board, (b) Coffee Board, (c) National Dairy Development Board (NDDB), and (d) Tobacco Board.

Problems in Agribusiness

- Availability and cost farm inputs
- Quality of agricultural produce
- Post harvest problems and value loss in Indian agriculture
- Large number of intermediaries
- Laws that force businesses to buy from monopoly markets at high prices, and not direct from farmers
- Poor roads, power, warehouses and cold storages, all of which increase the cost of farm produce
- Government price controls on many commodities, whether it is fertilizers or sugarcane or even wheat and rice.

Prospects of Agribusiness in India

Crating employment opportunities: There is considerable potential of creating more lucrative and attractive jobs for youth in agribusiness activities. Supermarkets and retail chains will need many times more young business graduates. It needs to be noted that already between 2000 and 2005 India created 1.1 million jobs per year, which is the highest among BRIC (Brazil, Russia, India, and China) countries.

Worlds 5th largest consumer by 2025: Indian incomes will almost triple over the next two decades and India will climb from its position 12th largest consumer market today to become the world's fifth largest consumer market by 2025, surpassing the size of Germany's consumer market.

India can be the world's largest food factory: India's food production today is equal to that of the USA and second only to China's. Food in India is already a $70 billion industry. By 2010 it will reach $170 billion. Value added foods, rising from $20 billion to $60 billion during this time. Growing food industry will raise agricultural yields, increase incomes, and create employment.

Sourcing is the best opportunity: Currently, the Indian food industry suffers from inefficiencies, wastage and value loss of the order of $14 billion. In order to develop basic food markets and reduce the number of intermediaries and the amount of wastage and inefficiency, it is essential to build more efficient procurement processes.

Government Initiatives in Indian Agribusiness

Increased domestic and foreign investment

- The government has abolished licensing for almost all food and agro-processing industries with some exceptions.
- Automatic investment approval, up to 100 per cent foreign equity or 100 per cent for NRIs and overseas corporate bodies is allowed for most of the food processing sector.
- 100 per cent FDI in fertiliser and pesticide sectors.

Information Technology

The Central and State governments, television, radio, internet, print media etc. can play a vital role, *e.g.* www.agmarknet.nic.in by giving agricultural market information to farmers and the Domestic and Export Market Intelligence Cell (DEMIC) in TNAU forecasting prices of agricultural commodity prices.

Export Promotion

The Exim Policy (2002–07) emphasizes on:

- Free export of most agricultural products.

- Setting-up of agri-export zones to enhance international market access, improve infrastructure facilities and to ensure better flow of credit.
- A new scheme has been introduced called the Vishesh Krishi Upaj Yojana (Special Agricultural Produce Scheme) for promoting the export of fruits, vegetables, flowers, minor forest produce and their value added products by increasing incentives for exporters of such products.

Credit

The government is planning to double credit to the rural sector over the next three years and also introducing immediate measures to ease the burden of debt and high interest rates on farm loans.

Crop Diversification

The government proposes to encourage farmers to diversify into areas such as horticulture, floriculture and oilseeds. **The National Horticulture Mission,** was announced in the budget 2004–05, and initiated in 2005–06 with an allocation of US $145 million for the mission.

Agri-business Consortium for small farmers: The Small Farmers Agri-business Consortium (SFAC) was set up in 1994. SFAC provides venture capital to projects and will be run, preferably by a banker, on purely business lines.

Initiatives of NGOs and Private Sector

Non-governments organizations (NGOs) are working with rural communities to establish agro-based micro-enterprises. **The private sector** is also active in this area; and multinational corporations are developing innovative marketing networks in rural areas by using information and communication technologies, **e-Choupal** is a good example initiated by ITC.

TNAU-agribusiness Model *(Directorate of Agribusiness Development (DAD))*

Goal: The Directorate of Agribusiness Development was started in February 2007 to promote agribusiness opportunities among agri-entrepreneurs and other stakeholders. The prime goal of this directorate is to establish a network of farmers, entrepreneurs, industry and public institutes and relevant stakeholders to promote agribusiness.

Activities

- Providing a portal of entry into the resources of the University for those seeking research and development services related to agribusiness.

- Organizing and conducting trainings/seminars on technology transfer and entrepreneur development.
- Commercialization of TNAU products and technologies.

Tamil Nadu Agricultural University has developed many viable agro-technologies and products. Some are already commercialized and some awaiting opportunities. Based on the recommendation of the Agri-Business Advisory Committee of this Directorate, the following products have been approved for commercialization as a first phase in the university are maize hybrid COH (M)5, tomato hybrid COTH 2, hybrid bhendi, TNAU coconut tonic, TNAU Panchagavya, TNAU Effective Microbial Extract (EME).

Efforts are being made to call for 'Expression of interest' by various reputed seed producing companies at national level.

Guidelines

- Non-exclusive basis
- Expression of interest at national level
- Company offering highest amount will be considered
- TNAU reserves all rights of continuing or rejecting the offer
- Company not allowed to grant sub-license
- Quality is monitored by TNAU
- Period of license is six years as per IPR guidelines
- All IPR rights of hybrids rest with TNAU
- It is mandatory to use the TNAU nomenclature.

Venture capital schemes: This is an innovative programme implemented in TNAU since May 2003 with the twin objectives of helping the farmers with provision of quality agricultural inputs (like seeds, seedlings, cuttings, grafts, bio-inoculants, vermin-compost, bio-fertilizers, bio-control agents, coconut tonic, advisory services etc.) and generation of revenue to the university. The seed money for the scheme is provided from the university education, research and development fund which has to be repaid within three years from the sanction of the scheme. As on date a total number of 138 schemes are functioning at various colleges, research stations and KVKs with a total budget outlay of Rs 168.036 lakhs. Out of the total seed money sanctioned, Rs 127.11 lakh have been repaid to the university. From the Venture Capital Schemes, a total amount of Rs 119.55 lakhs has been received towards the institutional charges for the years 2003–04 to 2007–08.

Guidelines

- The loan amount (seed money) ranges from Rs 0.5 to 3.0 lakhs

- Interest will not be charged for the sanctioned seed money
- The entire seed money is to be repaid within three years
- 25% on the total receipts earned shall be paid as 'institutional charges' after the payment of seed money
- The remaining 75% amount may be utilized for the next year's budget and infrastructure development

Offering Consultancy Services on Farm Products and Technologies

TNAU scientists offer consultancy services in various fields of agriculture and horticulture to industries, government and private firms and entrepreneurs. The mode of undertaking consultancy services are categorized into individual consultancy wherein the scientist concerned and the University share at 60:40 basis, and regarding the Institutional Consultancy the consultancy amount should be shared by the scientists (more than one scientist) and the University at 50:50 basis. So far, 167 consultancy projects offered from the year 2003 and the total amount of money Rs 56.7 lakhs received as University share.

Guidelines

- Client may contact the Individual scientist, HOD or university officer
- Consultancy services should be offered with the approval of the University
- Offering consultancy services by availing of eligible leave/loss of pay/ during holidays
- Concerned scientist should submit the completion report.

12 Strategies for Commercialization of Public - Bred Hybrids Through Private Sector Partnership in India

P.N. CHOWDHARY AND Y. YOGESWARA RAO*

ABSTRACT

After introduction of the liberalized "New Policy on Seed Development" 1988, the private sector seed companies expanded their research activities and some of them had collaborative arrangements with foreign companies. This has helped in free flow of germplasm material into the country from private companies abroad. To enable the farmers to have access to the best hybrids developed by ICAR / SAUs in different crops the proposed MOU / MOA between public and private sector organizations is discussed with numerous suggestions and justification to enhance "Public–Private Partnership" for marketing of hybrid developed by public sector.

Key words: Commercialization, new policy, partnership, public bred hybrids

Till the Government of India announced the "New Policy on Seed Development" in 1988, most of the varieties/hybrids released in different crops were from the public sector research organizations *i.e.* ICAR/SAUs. During the mid-sixties to late eighties, a good numbers of best performing hybrids in different crops were released by public sector organizations, which have become popular with the farmers in different states.

Few examples of the first generation hybrids are:

- Maize: Ganga-5, DHM 103,
- Sorghum: CSH-5, CSH-9
- Pearl millet: BK -560, BJ-104
- Sunflower: KBSH 1, APSH-11
- Cotton: Varalakshmi, NHH-44

After introduction of the liberalized "New Policy on Seed Development", the private sector seed companies expanded their research activities and

* Vikky's Agri-sciences Pvt. Ltd., Hyderabad - 500034.

some of them had collaborative arrangements with foreign companies. This has helped in free flow of germplasm material into the country from private companies abroad. In addition to this, CGIAR institutions like CIMMYT, IRRI, AVRDC have also provided the required germplasm lines to private companies. Till the early nineties ICAR/SAUs were supplying germplasm material liberally to private sector companies. Subsequently this facility was discontinued by many of the ICAR institutes and SAUs. Consequent to this, the private companies started utilizing the germplasm material provided by ICRISAT and few national institutes. The Private Sector seed companies are producing and marketing the hybrids of their own in many crops along with public bred hybrids like NHH-44 and CSH-9 in cotton and sorghum, respectively.

Growth of R&D in Private Sector Companies

With the introduction of liberalised new policy on seed development 1988, the Private sector seed companies have concentrated on developing their inhouse R&D facilities. The number of companies participating in crop improvement program has increased from a mere 5 companies in 1966 to 106 in 2004 as detailed below (Table 12.1).

Table 12.1. Number of companies with crop improvement programs (1966–2004)

Crop	Year			
	1966	1990	1998	2004
Pearl millet	2	12	18	25
Sorghum	2	10	12	15
Maize	1	06	15	20
Sunflower	–	08	17	12
Cotton	–	09	21	24
Hybrid rice	–	–	4	10

Source: K.R. Chopra "Challenges before the seed industry."

Table 12.2. Growth trends in the use of proprietary hybrids (1992–93 to 2002–03)

Crop	Percentage of proprietary hybrids (1992–93)	Percentage of proprietary hybrids (2002–03)
Sorghum	20	85
Pearl millet	50	90
Maize	33	93
Sunflower	90	98
Sorghum Sudan Grass	100	100
Cotton	27	78

Source: K.R. Chopra "Challenges before the seed industry."

Similarly the percentage of proprietary hybrids in rice also increased in the recent past and at present their market share is about 75% of the total hybrids marketed.

Even now the ICAR/SAUs are doing commendable job by providing the required breeder seed of public bred hybrids/varieties of different crops to the private and public sector organisations.

To enable the farmers to have access to the best hybrids developed by ICAR/SAUs in different crops it is very necessary to develop relationship between public and private sector organizations leading to "Public-Private Partnership."

The hybrids developed by ICAR/SAUs have to be marketed by the private seed companies under an MOU/MOA. This will help for continuous flow of location specific, biotic and abiotic stress tolerant hybrid seeds to the farming community which ultimately helps to increase productivity levels in different crops.

The MOU/MOA conditions must be liberal in trade related issues to avoid unhealthy competition among companies.

The MOU to be developed with the following objectives:

I. Sharing of breeding material and credit of IPRs with private sector seed industry:

- Breeding material developed by the ICAR/SAUs, *viz.*,
  - (a) Fully developed material (released and ready for release varieties/hybrids/parental lines)
  - (b) Breeding lines in the advanced stages of testing
  - (c) Raw germplasm and new introductions.

  can be shared with the private sector on non-exclusive basis.
- The terms for sharing of credit and IPRs on products developed under joint program with a given seed company shall be decided prior to undertaking such research programs through an MOU.
- The private companies can produce and market the public sector product by a popular name of it's choice.

II. **Consultancy services to be provided to the private sector**

Services of the expert scientists:

- Provision of short and long term consultancy including long term deputation of ICAR/SAU scientists to guide/assist private sector research and development programs to be considered.
- The scope of consultancy shall include project formulation, project implementation, project review, guidance and processing proposals to DBT etc. in case of transgenics.

- The Public Sector Institutes shall provide the services of a scientist of company's choice for consultancy and ensure that the same person's service is extended for repeated visits during the contract period.
- Problems of common interest to the industry will receive priority over those of individual companies.
- Both the parties shall respect the confidential nature of advice or information provided.
- Short term consultancy of one to two days for field visits for opinion seeking on field grow out test plots, guidance in seed production plots and to make evaluation of the hybrids/lines for tolerance to pests and diseases etc. to be provided.

Participation in technology transfer programs: The ICAR/SAU scientists to participate in technology transfer related programs such as field days, farmers meets etc. organized by the private sector, to enlighten farmers on the latest crop production technologies.

Training on seed production: Training to the personnel of private sector to be provided through short term courses on subjects of their interest which would include seed production technologies, DUS testing, IPR, hybrid breeding, maintenance breeding, molecular marker based screening /selection techniques, DNA fingerprinting, application of biotechnology tools for crop improvement, regulatory system for food/ environment safety etc.

The liberalized MOU terms of ICAR/SAUs will help the farmers to reap the benefits of the best hybrids developed by the public sector organizations.

III. Salient features of MOU suggested by ICAR

A. The LICENSOR will offer:

1. Genetically pure seeds of the parental lines of the hybrids agreed to be shared under MOU
2. Seed production technology and training, if needed at licensee's cost.
3. Monitoring facility of seed production plots on request. Specialists will monitor the hybrid seed production plots of the licensee. The cost for such visits (two visits in a cropping season) will be borne by the licensee.
4. Any other help, if required.

B. Responsibility of Licensee:

- The LICENSEE will not change the denomination of the hybrid, *i.e.* the name shall continue within or outside the country. The seed packet should indicate prominently the name of the hybrids assigned by the Licensor.
- The offer to "LICENSEE" is on non-exclusive basis.
- The "LICENSEE" will not use any of the lines (A and R) for developing a new hybrid without the prior approval of the "LICENSOR."
- LICENSEE shall duly acknowledge the receipt of parental lines from the Institute or its designated agent for production of hybrid seeds to be marketed by the former in consideration of which Licensee shall pay Licensor a royalty of 5% on the volume of commercial seeds sold.
- Such a royalty shall be payable once a year at the end of main selling season, *i.e.* 30th September every year. Licensor can have access to the relevant Licensee records, if it so desires, to ascertain the production and sale of the said hybrid, during the year.
- The License agreement shall be up to 3 years and its continuation shall be subject to review at the end of the contract period and on case-by-case basis and if both the parties mutually agree. The Licensor will be free to revoke the MOU unilaterally if the Licensee violates any of the clauses.
- The "LICENSEE" will recognize the "LICENSOR" as "INITIAL BREEDER" when developing ESSENTIALLY DERIVED VARIETIES (EDVs) from the parental lines.
- The "LICENSEE" will not sub-license the agreement material to a third party without prior approval of the "LICENSOR".
- The rules and regulations as mentioned in PPV & FR Act 2001 with respect of "EXTANT VARIETY/HYBRID" shall be acceptable to the "LICENSEE".
- These terms and conditions are subject to modification depending upon mutual agreement and developments in future with respect to IPR.
- The Licensee shall provide detailed production program and estimated quantity in advance to the Licensor.

All disputes arising out of this agreement shall be settled through mutual consultation of the parties. If the dispute is not so settled through mutual consultation by both the parties then the same shall be referred to the Director of the Institute who will be the final arbitrator and whose decision shall be final and binding on the parties.

IV. **Changes needed in MOU**

The Licensee may be permitted to market the seed outside the country by obtaining prior permission from the Licensor and following all the rules and regulations as per ICAR's IPR guidelines issued from time to time.

In case the Licensee wants to use any exclusive denomination (brand name) to trade the licensed hybrid the seed packet should contain a visible statement: "This hybrid is produced by using the parental lines licensed by ICAR/SAUs".

The Agreement shall be valid for a minimum period of 5 years initially from the date of signing the agreement and thereafter renewable for any further period on mutually agreed terms and conditions.

In case the licensee desires to further sub-license the seed of the said hybrid/variety for marketing, it is to be permitted. However the licensee will be responsible for payment of royalty for the quantities sold under such agreement.

Maintenance of Parental Lines and Production of Parental Lines

Maintenance of parental lines is to be done by the originating breeder of a public sector institute. The breeders of the private companies can be recognized as sponsored breeders and they can be entrusted with the job of seed production of parental lines A, B and R where large quantities of F_1 seeds are to be produced. However maintenance breeding has to be done by the originating breeder. The foundation seed is to be produced invariably by the private companies looking into their requirements.

The private companies have to inform well in advance the quantities of breeder/foundation seed proposed to be produced by them to the public sector organization.

Seed Production of Hybrids–Identification of Alternate Locations

Hybrid seed production (F_1) is to be taken up in congenial climatic areas where the per hectare seed yields will be high and also the synchronization of flowering of parental lines will be better. In addition to the location, the appropriate season and the period within the season for seed production are to be worked out. It is necessary to take up trial seed production in different locations, keeping in view particularly the temperature and relative humidity. For example, the hybrid rice seed production is mainly taken up in the districts of Northern Telangana and Rayalaseema regions of Andhra Pradesh during first fortnight of November for higher seed yields, keeping in view the temperature requirements. Suitable areas for seed production in different agro-climatic zones are also to be identified for this purpose. The seed produced must be available on time for the subsequent sowing season. The seed production must be economical to the farmers.

Economics of Hybrid Rice Seed Production in Andhra Pradesh

Table 12.3. Economics of rice seed production (Rs/ha)

Parameter	Variety (Fine)	Hybrid
Seed yield (Q./ha)	50	25
Cost/quintal (Rs)	900	3500
Value (Rs)	45,000	70,000
Rejected grain	2,811	5,250
Total return (Rs)	47,811	75,250
Cost of production (Rs)	30,850	39,750
Net return (Rs)	**16,961**	**35,500**

The hybrid rice seed producing farmer will get about 109% more income per hectare than the farmer producing varietal seed.

Popularization and Marketing of Hybrids

The marketing of hybrid seeds is to be taken up by the private and public sector seed companies and also by the Cooperatives. These organizations have to conduct the required number of field days and to give wide publicity about the hybrids and their performance in a specific segment/segments. The public sector scientists have to participate in these field days to disseminate the information on latest technologies in crop production for getting higher yields and economic advantage of the new hybrids. For the field days the farmers and the dealers have to be invited to show the benefits of the new hybrids.

Seed production and marketing is also to be encouraged through Cooperatives and Farmers Associations. There are many successful cooperatives involved in seed production and marketing in A.P. and other states.

Workshops on crop production, processing and value addition have to be conducted for farmers and farm women. This will help in transfer of latest technologies on crop production and value addition to the produce, which ultimately helps in increasing income of the farmers.

The single cross hybrids of maize are to be popularized on large scale for increasing productivity levels.

The importance of zero tillage for rabi maize for efficient utilization of residual moisture and for timely sowing is to be popularized. This concept is catching up in few Coastal and Telangana districts of A.P. This activity is to be taken up by Governmental agencies.

Impact on Food, Nutritional and Economic Security

By using the hybrid seed in place of varietal seed, the productivity level in different crops are increasing. As an example, the implementation of mini-mission in maize has contributed for increase in production and productivity in the country as given below:

Achievements under AMDP during Mission period from 1995–96 to 2003–04

- Increase in area 24.17%
- Increase in production 54.35%
- Increase in productivity 24.33%

Source: National Conference on Agriculture kharif Campaign 2005 GOI.

The increase in production and productivity of maize can mainly be attributed to the use of hybrid seed, and the thrust given by Governmental agencies.

Due to development and release of large number of location-specific hybrids in maize and availability of required quantity of seed, the farmers are able to get the seed at reasonable price.

There is a need for development of hybrids tolerant to biotic and abiotic stresses suitable for different sectors by both public and private sectors. There must be marketing thrust for popularizing the use of hybrids in remote areas.

Public-Private Partnership

Responsibilities of public sector research institutes and private- public sector seed companies and government organizations:

Public Sector Research Institutes (ICAR / SAUs)

- Sharing of germplasm material with private sector companies
- Production of nucleus and breeder seed
- To provide technical support for seed production to private and public sector seed companies
- Conducting training programs on seed production, DNA finger printing etc.

- Short-term consultancy to private and public sector seed companies on seed production technology, grow out tests and maintenance breeding etc.
- Screening of private hybrids (not included in ICAR/SAU trials) for pests, diseases and for grain quality
- To create coordination/net work cell
- Commercial wing to facilitate increased production of breeder seed.
- To form a consortium with seed agencies for research activities

Breeder seed production of company hybrids

Private companies, private and public sector seed agencies, and private companies/NSC/SSCs/Cooperatives)

- Foundation seed production
- Certified/truthful labeled seed production.
- To conduct demonstrations and field days on hybrids.
- To prepare and distribute publicity material such as pamphlets, posters, stickers etc.

State agricultural departments

- Extension activities on hybrid seed production and cultivation
- Identify farmers (men and women) for training in seed production and to conduct training programs
- Conducting demonstrations
- Publicity through audio-visual aids in FTCs etc.

13 Hybrid Seed Production of Emerging Field Crops—An Overview

S.K. Rao and Manpreet Hora*

ABSTRACT

India's seed industry has grown in size and level of performance over the past four decades. Both private and public sector companies/corporations are involved with the production of seed. Seed of the proprietary hybrids formed a significant portion of the total seed market. in terms of quantity, although proprietary hybrids had only 32.34% share of the market, in terms of value their share was 76%. Relying on hybrid technological rice has a very good prospects for commercial production and to play a key role in ensuring the future food security worldwide. The heritability of traits is important in improving the cross-pollination required in hybrid seed production of wheat are moderate to high, and thus reasonable genetic progress is expected in developing parents with improved seed yield efficiency. The CMS based hybrids will have a lot of scope to enhance wheat production and to play a major role for food security. The extensive testing of GMS based hybrids and limited evaluation of CMS based hybrids have demonstrated that high levels of exploitable heterosis is possible in pigeonpea. Genetic male sterility based hybrids in safflower promised nutritional security in India. Innovations in canola, through the introduction of herbicide tolerance, that reduced production costs showed promise in the mustard growing areas in the country.

Key words: Hybrid seed production, pigeonpea, rice, safflower, wheat

Hybrid breeding exploits the phenomenon of heterosis, which means the superiority of 'F_1' over either or both the parents. But, commercially, the standard heterosis is important, where the 'F_1' shows superiority over the standard check varieties. Because of this inherent advantage, hybrids in many crops are grown the world over. As an individual, a farmer produces

* Department of Plant Breeding and Genetics, Jawaharlal Nehru Krishi Vishwa Vidyalaya, Jabalpur - 482 004.

more per unit area, while as a country it increases its production and productivity, when compared to inbred varieties.

The advantage of hybrids or hybrid vigour was first reported by Koelreuter in 1673. since then many researchers have reported hybrid vigour in a large number of plant species. However, most of our understanding on heterosis comes from the work on maize. Beal studied the performance of inter-varietal hybrids between 1877 and 1882. He reported that certain F_1 hybrids yielded as much as 40% more than their parental varieties. Since then hybrids have been developed and cultivated in most crop species.The success of hybrid technology in any crop, largely depends on two essential prerequisites:

1. Distinct yield advantage compared to inbred varieties
2. Ability to produce hybrid seed on commercial scale at economic cost

RICE

There is need to go for hybrid rice because yield levels of semi-dwarf varieties of the green revolution era have plateaued. More and more rice to be produced on less land and with less input. Demand for rice is rapidly increasing with the increase in population, especially in less developed countries. Rice hybrids have shown 15–20% higher yield potential than inbred varieties under farmers' field conditions. Hybrids have shown their ability to perform better under adverse conditions of drought and salinity.

Rice is a strictly self-pollinated crop. Therefore, for developing commercial rice hybrids, use of male sterility is essential. Male sterility by genetic or non-genetic means makes the pollen unviable and such rice spikelets are incapable of setting seeds through selfing. Thus, a male sterile line can be used as female parent of a hybrid. A male sterile line, when grown side by side with a pollen parent in an isolated plot, can produce a bulk quantity of hybrid seed due to cross pollination with the adjoining fertile pollen parent. The seed set on male sterile plant is hybrid seed which is used for growing the commercial hybrid crop. This technology helped China to increase its rice production from 140 million tons in 1978 to 188 million tons in 1990.

Research at IRRI and in other countries indicates that hybrid rice technology offers opportunities for increasing rice varietal yields by 15–20% beyond those achievable with improved, semidwarf, inbred varieties. Adoption and success of hybrid rice technology will depend largely on practical seed production technology; economical seed yields from hybrid rice plots; and efficient national seed production, processing, certification, and distribution programs in public and private sector. Hybrid rice seed production technology involves specialized skills and requires a thorough understanding of various practices to minimize costs and maximize returns.

The successful commercialization of hybrid rice in many parts of the world is linked to the development of hybrid seed production technology. Although rice is a self-pollinated crop, a hybrid seed production mechanism has been developed using systems based on cytoplasmic and environmentally sensitive genetic male sterility. Chinese seed growers obtain on an average 2.5 mt/ha seed yields (range 1.5 to 7.4 mt/ha). Outside China, in tropical countries, mean seed yields are about 1.2 mt/ha (range 0.5 to 4.0 mt/ha). Hybrid seed production technology is economically viable in China and other countries with abundant cheap labor, because it requires about 50 man-days/ha more labor than normal rice cultivation.

The commercialization of hybrid rice technology in tropical countries has been facilitated by the research initiated in 1979 at the International Rice Research Institute (IRRI). It was soon recognized that Chinese rice hybrids were not suited to the tropics, and their grain quality was poor compared to the popular inbred high-yielding varieties grown in tropical countries. Therefore, IRRI researchers concentrated on developing suitable parental lines specially for the tropics, using the CMS system which had been found to be effective in China.

Within a decade, some commercially usable CMS and restorer lines and some elite hybrids were identified and shared with NARS for evaluation and utilization. Some of these hybrids were released for commercial cultivation in the Philippines (IR64616H, IR68284H), India (IR64611H, IR64618H, IR69690H), Vietnam (IR64615H, IR64616H, IR69690H) during 1993–98. Some IRRI CMS lines, IR58025A and IR62829A, were also used by national programs to develop local hybrids in India (*e.g.* APHR-1, APHR-2, CORH 2, KRH 2, ADTRH 1, etc). Concurrently, several private seed companies in Asia were also provided with IRRI-bred parental lines.

Hybrid seed production technology involving these male sterility systems is complex and labor-intensive. It also requires a sophisticated seed industry infrastructure. Hybrid rice seed production technology is different from that for inbred rice seed production, and is more complex than seed production for many other hybrid crops. Currently, the most popular male sterility system is the CMS.

Constraints

1. Non-synchronization of flowering of male and female lines
2. Poor grain quality characteristics of hybrid rice which is not preferred by the consumers, generally, after cooking the taste of cooked rice is not as palatable as non-hybrid rice varieties, therefore, the farmers are not getting better price in the market
3. High cost of hybrid seed production
4. Higher percentage of chaffy grains in the ears of hybrid

5. Lack of proper seed production technology to increase hybrid seed yield
6. For full exertion of panicles, spray of gibberellic acid is essential resulting in high cost of seed production
7. Susceptibility to some of the major pests and diseases
8. Yield increase in hybrid rice is in the range of about 15–20% but generally, farmer's expectation are very high say 30–40% as it is in the case of cross pollinated crops such as maize, bajra sorghum and cotton. Therefore, many farmers are not adopting hybrid rice cultivation
9. Lack of extension activities to bring awareness among the farmers about the hybrid rice cultivation technology.

For hybrid rice seed production to succeed, a sufficient quantity of pollen grains must be deposited on the stigma lobes of each spikelet of the seed (male sterile) parent. It helps if the pollen parent grows to a greater height than the seed parent. Other plant characteristics which influence this include small and horizontal flag leaves, the number of panicles per square meter, the number of spikelets per panicle, good panicle exsertion, and synchronized flowering of seed and pollen parents. The floral traits influencing outcrossing in rice include: stigma size, style length, stigma exsertion (in the seed parent) and anther length, filament length, and pollen number/anther (in pollen parent). The flowering behavior traits influencing outcrossing in rice are: number of days of blooming, time of blooming, duration of blooming, duration of floret opening, and angle of floret opening. The environmental factors influencing outcrossing in rice include temperature, relative humidity, light intensity and wind velocity. In China, conditions favorable for good outcrossing in rice have been identified as: a daily temperature of 24–28°C, a relative humidity of 70–80%, a diuranal difference in temperature 8–10°C and sunny days with a breeze. Suitable field conditions include: fertile soil, a dependable irrigation and drainage system and a low risk of disease and insect infestations.

Seed yield from a male sterile line used in hybrid rice seed production is a function of three factors. First is the yielding ability of the male sterile line, as determined by the yielding ability of its maintainer, the fertile counterpart. Second is the proportion of the male sterile line to the pollen parent. The third is the outcrossing rate of the male sterile line. Improvement in any of these three factors can help to increase the seed yields of hybrid rice. Extensive research has led to the identification of the following guidelines for successful hybrid rice seed production.

Isolation: A field should be used which is isolated from other rice varieties by sufficient distance (100 m), time of flowering (at least 3 weeks) or some natural or artificial barrier. The isolation distance can be decreased by increasing the number of pollen parent rows around the seed production plot. Barrier isolation can be effected through topographical features, or by the use of tall plants or artificial obstacles to block pollen movement.

Timing of seeding of the parent lines: The two parental lines should be sown and transplanted at the right times so that their flowering is synchronized. This is the most critical step in hybrid seed production, and is practiced by using one of the three methods described below:

The seeding date of the two parents can be determined according to the difference in the time they take to reach maturity. The parent which matures earlier must be sown later than the other parent.

The seeding date of the two parents can be determined by the difference in the number of leaves at the main culm. If this trait is stable, the date of seeding of the early heading parent can be decided according to the predetermined number of leaves present on the later heading parent.

The differential seeding of parental lines can be based on their effective accumulated temperature (EAT) from seeding to flowering. The EAT of a particular variety is calculated by summing up the mean temperature of days with a range within 12°C to 27°C. The parent which has a higher EAT is sown earlier, and the other parent is sown after the difference in EAT between the two has accumulated.

Optimum ratio of seed parent : pollen parent plants: The optimum seed parent : pollen parent ratio must be used according to the spikelet number per panicle of the parental lines, and the quantum of the pollen load of the pollen parent. The seed parent is planted densely, to increase the number of spikelets per square meter. The pollen parent is more widely spaced, to promote tillering and help ensure that the pollen load is greater over a longer period.

The seed production plot must be given good cultural management. Weeds, diseases and insects must be controlled, and the optimum fertilizer rate must be applied. If the panicle initiation stage of the two parental lines is found to occur at different times, they should be given different application rates of nitrogen, phosphate and potassium fertilizer, to adjust the flowering date of the parental lines by 2–5 days.

If the parental lines have long, erect flag leaves which might obstruct the movement of pollen, the flag leaves should be clipped by one-half or two-thirds of their length, one to two days before the initial heading. However, this practice has a negative effect that it may spread diseases such as bacterial blight, and sheath blight through the wounds inflicted on the leaves. Besides, this practice also reduces photosynthetic activity. Parental lines possessing short narrow, flag leaves do not need to be clipped.

GA_3 should be sprayed on parental lines which show poor panicle exsertion. Besides improving panicle exsertion, GA_3 also increases the stigma exsertion rate, as well as the time during which the spikelet is open and the stigma is receptive. Differential application of GA_3 also increases the relative height of the pollen parent over the seed parent. GA_3 is generally sprayed using a knapsack sprayer. Spraying begins when 5–10% of the rice plants are at the heading stage, and lasts for 2–4 consecutive days.

Spraying should be done on a calm sunny day, usually in the afternoon. It should not be carried out if rain is expected within 24 hours.

Hybrid rice seed producers in China use a very high dosage (150–300 gm/ha) to get high seed yields. However, outside China the high cost of GA_3 limits seed growers to using only 45–50 g/ha.

Roguing of off-types should be carried out at the stage of the stage of maximum tillering, flowering and maturity stages, to prevent any cross-pollination between them and the true-to-type seed and pollen parent plants.

Pollen and seed parents should be harvested, threshed, dried and cleaned separately and carefully, to avoid any mechanical mixture of the seeds of parental lines.

Quality control for hybrid rice seed: Quality control is vital to a widespread adoption of hybrid rice. Farmers who buy expensive hybrid seeds expect high quality. Quality control is carried out throughout the whole process of seed production. It includes the nucleus, breeder, foundation and certified seed production of both parental lines and F_1 hybrids.

Seed quality standards are met by intensive roguing of off-types (from seeding to harvesting) and careful handling during harvesting, threshing, drying, cleaning, processing, bagging and labeling. Grow-out tests, germination tests, seed vigor tests etc. should also be done before the seeds are sold commercially.

Economic viability of hybrid seed production: The availability of quality seed at a reasonable price is a crucial factor behind the large-scale adoption of hybrid technology for any crop. Chinese success in large-scale adoption of hybrid rice can also be attributed to the efficient and economical hybrid seed production and distribution system, organized by the state-owned seed industry.

Hybrid seed production technology is considered labor- and knowledge-intensive. It involves various risks, especially in the early stages when seed producers are still lacking experience. Typical problems are poor synchronization of the parental lines, and unfavorable weather conditions.

More encouragingly, good results have been achieved in developing super hybrid rice varieties since the initiation of the super rice research program in 1996. Several pioneer super hybrids have a yield advantage of around 20% over current three-line hybrids on commercial scale. The area planted to super hybrid rice was 240,000 ha and the average yield was 9.6 t/ha in 2000. The area under super hybrid rice increased to 1.4 million ha with an average yield of 9.1 t/ha in 2002. In addition, a two-line super hybrid P64S/E32 and a three-line super hybrid II-32 A/Ming 86 created a record yield of 17.1 t/ha in 1999 and 17.95 t/ha in 2001, respectively. In the meantime, the grain quality of the super hybrid rice varieties is very good. Now efforts are focused on developing second generation super hybrid rice. Its yield target is 12 t/ha on large scale, and good progress has been made. Therefore, the super hybrid rice shows a very bright future. If super hybrid rice covers an annual area of 13 million ha in China and calculating by a yield increase of 2.25 t/ha, it is expected that the annual increased grains will reach 30 millions tons, which means 75 million people more can be fed every year. Hybrid rice has been proved to be a very effective approach to greatly increase yield not only in China, but also outside China. The ever-forwarding technology improvement by scientists, its dissemination by seed industries and extension workers and the policy and financial support by national and local governments contribute greatly to the success in development and use of hybrid rice technology. And We firmly believe that hybrid rice, relying on scientific and technological advances and the efforts from all other aspects, particularly from FAO and IRRI, will have a very good prospect for commercial production and continue to play a key role in ensuring the future food security worldwide in the new century.

Development of high yielding hybrids is one of the major objectives. During the last 10 years more than 1000 experimental hybrids developed by different network centers, IRRI and private sector seed companies have been evaluated in multi-location trials. Of these, 13 public bred hybrids showing consistent yield superiority over local inbred check varieties have been released for commercial cultivation in different regions by the respective state variety release committees

Out of the 16 hybrids released in India so far, large scale seed production of five hybrids, *viz.*, DRRH-1, KRH-2, Sahyadri, PHB-71 and PA 6201 has been taken up by the public and private sector seed agencies. Hence, seed of above mentioned hybrids is available to the rice farmers for large scale cultivation.

At present hybrid rice is reported to be grown on approximately 200,000 hectares. Area under hybrid rice will further increase after heterotic hybrids suitable for high productivity areas of Punjab, Haryana, coastal region of Andhra Pradesh and shallow lowland areas are identified and an effective transfer of technology programme is taken up vigorously in the target states.

Several agronomic trials were conducted to standardize the cultivation package for the rice hybrids. They involved finding out the optimum seed rate, seeding density in the nursery, number of seedlings/hill, suitable planting dates, planting geometry and nitrogen requirement etc. Seed is the major costly input in hybrid rice cultivation. Therefore, efforts have been made to reduce the seed rate without adversely affecting grain yields. All the released hybrids in India and the promising pre-released hybrids have moderately acceptable quality but they cannot be compared with some high quality varieties such as Samba Mahsuri in Andhra Pradesh, White Ponni in Tamil Nadu and Basmati varieties of northern India. Separate breeding programmes have to be initiated to develop hybrids of very high quality. Hybrids have been evaluated for quality characteristics. Some of the hybrids, *viz*., ADTRH-1 and DRRH-1 possess good quality characters. With the availability of a large number of CMS lines and the pollen parents, it would be possible to develop hybrids with desired quality characters.

The success of hybrid rice technology depends on the efficient and economic hybrid seed production, so as to provide good quality seed at reasonable price to the farmers. Hybrid seed production, in self pollinated crops like rice is quite tedious. Therefore, extensive trials were conducted to standardize the seed production package. Various requirements for successful seed production such as ideal locations/seasons, seed rate, row ratio, GA_3, supplementary pollination techniques etc., have been worked out and a seed production package has been developed. The details of the package are as under.

Seed producers can obtain seed yields of about 1.0 to 1.5 t /ha of hybrid seed by adopting this package. There is a need to refine the present technology so as to enhance the seed yields up to 2.0 to 2.5 t/ha so that the seed cost can be reduced.

India has a very well organized seed industry both in public and private sectors and this can be effectively utilized for hybrid rice seed production. Realizing fully well, the potential of seed industry, enough encouragement was given to the seed sectors to take up hybrid seed production. Required parental lines were made available to the interested companies and their personnel were imparted training on hybrid rice seed production technology. This step has really helped many companies and some of them have established full fledged R&D facilities for hybrid rice. Now the seed industry has fully geared to meet the seed requirement and what it needs is the demand which has to be created by well organized transfer of technology activities.

The leading private sector seed companies in hybrid rice seed production are Hybrid Rice International, PHI, Mahyco, Indo-American, Parry Monsanto, Advanta India, Syngenta and Nath Seed company. Half a dozen other private seed companies are also engaged in hybrid rice seed production on a smaller scale.

Among public sector seed agencies, National Seed Corporation (NSC), State Farms Corporation of India (SFCI), Andhra Pradesh State Seed

Development Corporation (APSSDC), Karnataka State Seed Development Corporation (KSSDC), West Bengal State Seed Development Corporation (WBSSDC), and U.P. Terai Seed Development Corporation (UPTSDC) are presently engaged in hybrid rice seed production on much smaller scale. All these public sector seed agencies put together, produce around 100 tons of hybrid rice seed annually.

Hybrid rice is likely to become popular and adopted in large areas during the next decade. Huge quantities of hybrid rice seed will be needed. Thus there is ample scope for all the private and public sector seed agencies in India to produce large quantities of hybrid seed in the years ahead to popularize rice hybrids.

WHEAT

Interest in the development of hybrid wheat was initiated by the discovery of cytoplasmic male sterility in wheat for the first time by Kihara (1951). He found the CMS system in the back cross derivatives of the cross between *Aegilops caudata* and *Triticum aestivum* developed through transfer of bread wheat genome in *Aegilops* cytoplasm through repeated back crossing with *Triticum aestivum*. Fukusawa (1953) found similar system in the back cross derivatives of the cross *Aegilops ovata x T. durum* where *T. durum* genome was completely reconstituted in the *Aegilops* cytoplasm. Later on different sources of cytoplasmic sterility and male fertility restoration were found in the U.S.A. (Wilson and Ross, 1962). Brigale (1963) indicated positive heterosis for grain yield ranging over 50–100% in wheat. In the last 15 years significant efforts have been made for commercial exploitation of hybrid wheat through the use of gametocide. Today France, Italy and USA have developed commercial wheat hybrids through CHA route.

In India efforts on hybrid wheat were initiated in the sixties following CMS approach, however, no significant results were obtained. In 1995, Directorate of Wheat Research, Karnal along with its cooperative centres decided to readdress hybrid wheat with an emphasis on CHA approach. The pre-breeding approaches are targeted on long-term basis to identify and develop parents for specific hybrids with commercial heterosis. Since 1995, the Directorate of Wheat Research and its cooperative centres have made significant headway in this direction. The hybrid wheat research was addressed primarily at following fronts:

- Formulate, evaluate and operationalise chemical hybridizing agents in producing hybrids.
- Develop reliable CMS and restorer systems.
- Investigate floral biology and related seed borne diseases.

Research and development of hybrid wheat has a long history. Pickett (1993) has thoroughly summarized the principal events in this history through 1990. He begins the chronology with the proposal for hybrid maize by G.H. Shull in 1909 and ends in 1990 with Cargill ceasing development of hybrid wheat in the United States, yet continuing with commercialization in Australia and Argentina.

Additional developments in the chronology of hybrid wheat would have to include advances in the development of the science of genomics, which could lead to an improved understanding and ability to exploit the phenomenon of heterosis in this complex allohexaploid. Transformation of wheat, now possible with particle gun and *Agrobacterium* sp. technologies, also may lead to further investment in hybrid wheat due to genetic dominance of transformed genes and an ability to capture the value in a hybrid crop.

Cytoplasmic male sterility: Wilson and Ross (1962) were the first to describe a workable cytoplasmic male sterile, nuclear-restoration system when they reported that *Triticum timopheevii* cytoplasm induces male sterility with largely neutral effects on agronomic and quality characteristics. This discovery led to the establishment of a number of research programmes to evaluate alternate cytoplasmic sources of male sterility with common wheat and the nuclear genes necessary to restore fertility to the F_1 hybrid.

Virmani and Edwards (1983), in a comprehensive review of cytoplasmic male sterility, reported that 15 different cytoplasms of the genera *Triticum* and *Aegilops* have been recognized and that several of these could form the basis of a CMS system. However, they also stated that the *T. timopheevii* system has gained widespread use due to deleterious effects of the other cytoplasms on various traits and/or because no advantage existed over the *T. timopheevii* system.

Development of restorer lines (R-line) to effect restoration of normal pollen fertility to the F_1 hybrid involves the discovery and incorporation of fertility restoration (*Rf*) genes. Restoration genes from a number of species and from common wheat have been identified and are scattered throughout the ABD genomes (Wilson, 1984). The species that is the cytoplasmic donor is often a source of fertility restoring genes, and a number of *Rf* genes have been derived from *T. timopheevii* (Virmani and Edwards, 1983). Fertility restoration is under complex genetic control, and is further complicated by differences among females for ease of restoration and by the effect of environment on fertility restoration (Wilson, 1984). Wilson (1968) attempted to classify environments as "shallow sterile", "sterile" and "deeply sterile". All of this has caused much difficulty in developing agronomically acceptable R-lines that give complete fertility restoration of the F_1 hybrid.

The CMS system requires complete and stable male sterility and complete restoration of fertility to the F_1 hybrid in a number of environments. While R-lines have been developed that restore complete fertility, the difficult task of continuing to make genetic progress in developing agronomically improved R-lines remains as the most significant obstacle in the continued development of CMS hybrid wheat. The use of marker-assisted selection for *Rf* genes, as with restriction fragment length polymorphism (RFLP), would seem to increase the probability of making agronomic progress while retaining *Rf* genes in the male population (Ma and Sorrells, 1995).

It should be pointed out, however, that some of these disadvantages are also issues in a CMS system. Genotypes x environment interactions are, for example, also a problem in developing males with adequate fertility restoration in an array of environments. Similarly, male fertility can be a problem with a CMS female that does not have stable cytoplasmic sterility or that has a B-line that cannot be converted to complete pollen sterility (Wilson, 1984).

Heterosis: Briggle (1963) was among the first to review heterosis in wheat. Varying estimates of heterosis were reported. Pickett (1993) presents a thorough review of the North American and European large- and small-scale trials investigating the magnitude of heterosis for grain yield and other traits. In the small-scale trials, sowing densities were low or were space planted. The large-scale trials more closely approximated normal sowing densities.

The small-scale European trials of winter wheat hybrids reported from 1934 through 1976 appear to indicate that heterosis of 30 per cent and greater is possible. Eleven authors reported maximum high-parent heterosis ranging from 14.2 to 74 per cent. However, Pickett (1993) holds their validity in question due to the low sowing density. The results of three authors reporting on heterosis in spring wheat hybrids were also summarized: (i) heterosis was found in two of the six hybrids, the highest being 7.5 per cent above the high parent; (ii) forty-four of 48 hybrids showed heterosis, with 40.7 per cent being the maximum; and (iii) no hybrids were detected that exceeded the best available commercial varieties.

Heterosis in winter wheat hybrids evaluated in small-scale North American trials is summarized by Pickett (1993) as "clearly demonstrating simple heterosis in winter wheat". Maximum better-parent heterosis ranged from 24.7 to 76.0 per cent among the hybrids reported on by 13 authors surveyed from 1931 to 1982. In small-scale spring wheat trials conducted from 1921 to 1980 in North America, Pickett (1993) concludes that a number of reports failed to confirm a statistically significant level of heterosis for grain yield. However, maximum high-parent heterosis for the best hybrid in the respective trial reported by each of the 12 different authors still ranged from 3.9 to 88.0 percent.

Pickett (1993) also compiled results on heterosis in spring wheat hybrids from authors in India, Pakistan, Bangladesh, Australia and Mexico.

Impressive levels of heterosis were reported, although it must be noted that a number of measurements were based on single plants. Maximum high-parent heterosis from 1938 to 1986 reported by 13 authors ranged from 9.0 to 160.4 per cent.

Economics: The successful adoption of any hybrid crop requires that a certain level of economic profitability must be met. Wheat—historically cultivated as a self-pollinated crop—requires a level of economic justification similar to other hybrid crops, much like corn and sorghum, if it is to be widely adopted. For the present time, hybrid seed costs more than certified varietal seed or farmer-saved seed.

In areas that traditionally have higher seeding rates (120 kg/ha and greater), a correspondingly greater increase in hybrid yield is required to demonstrate an economic advantage for hybrids. Areas with moderate (55 to 85 kg/ha) seeding rates can have a lower hybrid advantage and still prove economical. Environments with low yield potential, such as extreme dryland ecosystems, may not be able to deliver an economic hybrid advantage despite relatively low seeding rates. Limited supplies of hybrid wheat seed currently are being sold in the United States in the range of US$ 0.65 to 1.10/kg, compared to a certified seed price range of approximately US$ 0.25 to 0.55/kg. An estimate of the true cost of farmer-saved seed in the United States is US$ 0.17 to 0.35/kg.

Commodity price: Higher commodity prices make the cost of hybrid seed easier to justify. When these factors are present in the marketplace, they need to be considered in an economic analysis of the benefits of hybrid wheat. This indicates that the economics are becoming sufficiently attractive for growers to risk planting hybrid wheat on a portion on their farm to determine economic viability using their own management practices. Hybrid wheat represents a sustainable technology for keeping pace with this increasing demand from diminishing resources.

Developing breeding populations: A number of authors have noted the difficulty of adequately evaluating hybrid performance when seed quantities are small. When only limited numbers of crosses can be evaluated, it is difficult to sample a large number of parents and select those that combine well and have the highest heterosis. One of the advantages of an effective CHA is the ability to produce large amounts of seed of a large number of hybrids, thus sampling a large number of breeding lines or populations in the breeding programme.

Selection of efficient females and males: The development and improvement of heterotic breeding populations is important in hybrid wheat breeding. However, since wheat has evolved as a self-pollinated species, it is equally important that these populations and inbreds be developed for out-crossing efficiency. This point needs to be clearly stressed, as seed yields in the hybrid production field are important to the economic viability of the hybrid seed enterprise. It has long been recognized that a major constraint in the development of hybrid wheat has been the low seed-multiplication ratio in production fields and the relatively high seeding rates required for commercial crops.

Floral biology of the wheat plant, including heritability estimates for the various floral attributes affecting cross-pollination, is well summarized by Pickett (1993) and Virmani and Edwards (1983).

Information on anther extrusion, anther length and duration of pollination in males is readily obtainable. Flower opening is easily observed in male-sterile females, but other traits, such as stigma size or pollen viability, are more difficult, though not impossible, to evaluate, based on the authors' experience.

Seed production: A number of issues are important in the economic production of hybrid seed. These include the identification of an optimal production environment, selection and training of growers to manage and produce the hybrid seed, and identification and management of factors influencing the relative flowering dates of the female and male parents.

Environment: Selection of a production location will have a significant impact on both the stability and the yield of hybrid seed and the resulting cost of hybrid seed. Experience has shown that lower costs of hybrid seed production generally will be obtained under high yield conditions associated with irrigated areas and reliable wind patterns during the pollinating season. Currently, the best locations identified for hybrid seed production in the United States are in the western high plains of the hard winter wheat region and in the Pacific Northwest.

Maximum hybrid seed set for a given hybrid combination will occur when the lemma and palea in the female plant and the open stigma receptive at the same time the male is shedding pollen. This means that the male should be two to five days later in days to heading than the female to achieve optimal when flowering coincidence (nick). Flowering data, including the day of 50 per cent pollen shed (male) and 50 per cent flower opening (female), and seed yield data were collected in top-cross nurseries over nine years in the HybriTech US soft wheat programme. Nick was calculated as the day of 50 per cent pollen shed minus the day of 50 per cent flower opening. Hybrid seed yields were optimized when flowering nick was plus two or three days, although commercially viable seed yields with wider or narrower nicks have been obtained.

Females are often planted at higher populations than males to reduce tillering and minimize differences in growth stage for optimal efficacy of the CHA.

One of the potential benefits of molecular genetic hybridization systems would be complete elimination of separate male and female strips. This would greatly simplify planting and harvesting operations, and should result in increased hybrid seed production and a lower cost of goods due to the close juxtaposition of male pollinators to receptive females. Such a system would also eliminate much of the risk associated with unfavourable weather at the time of CHA application, as discussed below.

Quality of hybrid seed from the production field may be compromised if sterility of the female parent is not complete. Cleaning and conditioning of hybrid seed is essentially the same as would be done for a certified crop of varietal wheat seed. Large seed size, while generally favourable to seedling vigour and growth, can add to the cost of goods and reduce the economic advantage of hybrids if planting high populations is the prevalent farm practice.

PIGEONPEA

Pigeonpea ranks sixth in area and production in the world in comparison to other grain legumes such as beans, peas, and chickpeas, it is used in more diverse ways than others It is now widely grown in the Indian subcontinent that accounts for almost 90% of the world's crops. Pigeonpea is an important crop of rainfed tropics of Asia, Africa, Latin America and the Caribbean. In India, it is predominantly grown in the states of Maharashtra, Uttar Pradesh, Karnataka, Andhra Pradesh, Bihar and Gujarat with a total area over 3.8 million hectares, but the country is still short in grains by a margin of about 500,000 tons annually. One of the main reasons for the shortage is its low mean productivity (600–700 kg ha^{-1}), which is a matter of concern not only to the farming communities but also to scientists, policy makers, and other stakeholders.

ICRISAT and its national partners are hopeful that new pigeonpea hybrids will soon make those concerns a thing of the past. Since ICRISAT has the global mandate for genetic improvement of pigeonpea, its crop improvement team took challenge of enhancing the crop productivity through the development of commercial hybrids, which is a well-exploited and proven technology in cereals and vegetable crops. After 30 years of concerted efforts in hybrid pigeonpea breeding research at ICRISAT, an efficient hybrid technology (the first in any food legume) based on cytoplasmic- nuclear male-sterility (CMS) system was developed.

Hybrid Technology

- To increase the crop's productivity, hybrid technology was developed by ICRISAT with its partners and the world's first pigeonpea hybrid ICPH 8 was released in 1991.
- Cytoplasmic nuclear male-sterility (CMS) systems have been developed at ICRISAT using three wild relatives of pigeonpea.

- A large number of diverse seed parents and fertility restorers with important agronomic traits including disease resistance have been developed.
- New CMS-based hybrids, exhibiting 20–30% yield advantage over pure line cultivars have been identified.
- The CMS-based hybrid technology has been shared with six private sector seed companies and public sector institutions.

Early Research: The development of hybrid technology in pigeonpea was initiated with the discovery of two sources of genetic male-sterility (GMS) from germplasm The expression of male sterility is controlled by a single recessive gene. These male-sterile have prominent anther size, shape and color, which provide an effective and easy way of identifying the male sterile plants in the field before anthesis. ICRISAT and its research partners in India have extensively used these sources in hybrid breeding programs. The following high yielding GMS-based pigeonpea hybrids were released for commercial cultivation.

ICPH 8: This was the first pigeonpea hybrid released in the world. Since no commercial hybrid was ever bred in any food legume, the release of ICPH 8 by ICRISAT in 1991 is rightly considered a milestone in the history of breeding food legumes. Evaluations from 100 yield trials showed that ICPH 8 was superior to controls UPAS 120 and Manak by 30–34%. In the on-farm trials also ICPH 8 maintained its superiority over the controls by more or less similar grain yield margins (Saxena *et al.*, 1992).

PPH 4: In 1993, Punjab Agricultural University, Ludhiana, India, developed another pigeonpea hybrid, PPH 4. This hybrid was found suitable for pigeonpea-wheat rotation and the yield outdid that of the control by a margin of 14%.

CoH 1 and CoH 2: In 1994, a short-duration hybrid, IPH 732, was released by Tamil Nadu Agricultural University (TNAU), Coimbatore, India, as CoH 1 which recorded 32% higher yield over control (Murugarajendran *et al.*, 1995). In 1997, TNAU released another pigeonpea hybrid CoH 1 with 13% yield superiority over Co 5.

AKPH 4104 and AKPH 2022: Two pigeonpea hybrids, AKPH 4104 (in 1997) and AKPH 2022 (in 1998), were released by Punjabrao Krishi Vidyapeeth, Akola, India. AKPH 4104, is a short duration hybrid with 64% superiority over the control UPAS 120 while AKPH 2022 is a medium-duration hybrid with 35% superiority over BDN 2.

World's first CMS based pigeopea hybrid: The first CMS based pigeonpea hybrid, GTH-1 developed at GAU, S. K. Nagar has been identified and released for cultivation in Gujarat state. Parents of this hybrid are GT 288A (CMS line) and GTR-11 (restorer line). Based on yield trail (2000–

2003), it gave 32% yield superiority (1627 kg/ha) as compared to the best local check GT 101 (1288 kg/ha). This is a short duration (140 days) hybrid possessing indeterminate plant type and large white seeds. This hybrid was notified in 2007 and is expected to increase pigeonpea production substantially.

Hybrid pigeonpea: breaking the yield barrier: Scientists working to improve the grain legume pigeonpea–*Cajanus cajan* (L.) Millsp.– announced recently that the new hybrid ICPH 2671 produces nearly 50 percent more grain than the popular Indian cultivar Maruti, convincingly breaking the yield barrier that has had them and farmers frustrated for many years.

Development of the world's first commercially viable system for producing hybrid pigeonpea seed was completed 2 years ago by the International Crops Research Institute for the Semi-arid Tropics (ICRISAT), working in close collaboration with the Indian Council of Agricultural Research (ICAR). A public good resulting from more than two decades of sometimes frustrating research, the technology represents "a major breakthrough," says ICRISAT Director General William Dar, permitting pigeonpea yields of 3 to 4 tons per hectare.

Eminent agricultural scientist M.S. Swaminathan predicts that the new pigeonpea hybrids, with their "quantum leap in yield," could open the way for a revolution in the production of this important pulse, similar to the transformation of wheat and rice production made possible several decades ago by novel semi-dwarf varieties.

According to K.B. Saxena, the ICRISAT scientist who led the development of development of the new pigeonpea technology, 100 to 150 tons of hybrid seed, enough to plant about 25,000 hectares, should be available to farmers in 2008. It is being produced with the help of 16 public and private seed companies.

Finding this solution was highly complex, requiring years of painstaking effort and skilled use of a pigeonpea wild relative (*Cajanus cajanifolius*), collected from the forests of Madhya Pradesh. The hard-won result consisted of cytoplasmic nuclear male sterile parents with high enough seed yields to permit viable commercial production of pigeonpea hybrids. In the 2 years since this system was devised, ICRISAT has tested more than 300 experimental hybrids, finding ICPH 2671 to be the most outstanding.

Equitable access: While the pigeonpea hybrids offer a sizable benefit, there is also a cost. Rather than produce their own pigeonpea seed year after year, farmers will need to obtain new supplies of certified hybrid seed each year, just as farmers do with hybrid maize and rice. ICRISAT is working closely with a consortium of private- and public-sector seed companies in India to commercialize pigeonpea hybrids and ensure that ample supplies can be made widely available within the next couple of years. Researchers in the Philippines, Myanmar and China are also using the new cytoplasmic nuclear male-sterile system to develop pigeonpea hybrids for their countries.

Natural out-crossing: The efficiency of any commercial hybrid seed production programme depends on the effectiveness of controlled natural mass pollen transfer mechanism from male parent to female parent. This ensures quality hybrid seed production economically. In practice, it is achieved by developing a productive male sterile line which is devoid of pollen grains, and by fertilizing its flowers with the pollen of designated male fertile plants by natural means. In pigeonpea, the natural mass transfer of pollen grains is accomplished by a number of pollinating insects.

In pigeonpea, most germplasm accessions have normal flowers and these lines are prone to natural outcrossing, which was always considered a major bottleneck in maintaining genetic purity of pigeonpea cultivars and selections until ICRISAT scientists utilized it for the genetic enhancement of yield through hybrid breeding.

Isolation distance: Since natural outcrossing in pigeonpea is effected by insects, safe isolation distance is essential to produce quality seed of hybrid and their parental lines. In India, the seed certification standards for isolation distance for pigeonpea inbred varieties are 200 m for breeder seed and 100 m for both foundation seed and certified seed. In pigeonpea, so far there is no reported studies on the requirement of isolation distance for CMS based hybrid seed production.

Cost of hybrid seed production: So far no studies have been carried out on the cost of hybrid seed production of CMS based hybrids. However, some inferences about the cost of hybrid seed production could be drawn using the data generated earlier with the GMS based hybrid technology and this information is reviewed here. The initial seed production feasibility study, jointly conducted by ICRISAT Patancheru and the Maharashtra Hybrid Seed Company (MAHYCO) in 1979–1980, showed that hybrid pigeonpea seed could be produced at a very reasonable cost. In these studies roguing in the female parent was found to be the most critical activity, and it accounted for a major proportion of the production cost. In the CMS based hybrid technology the roguing operation is totally eliminated and hence the production cost of both the female parent and hybrid will be reduced further by a significant margin.

Enhanced production through hybrid pigeonpea: The new hybrid named ICPH 2671 will yield 40 to 50 per cent more yield (that is 3–4 tonnes per hectare) than the other commercial varieties and offers a hope of pulse revolution in India and other developing countries. It is ideally suited for growing in drylands and water scarce regions and is highly resistant to two major diseases-fusarium wilt and sterility mosaic-prevalent in all the major pigeonpea growing areas. The ICRISAT hybrid technology is likely to increase the agricultural productivity and farmers' income in drylands. In spite of many agricultural research institutions releasing several varieties in the past five decades, the productivity of red gram has remained low, and this technology promises to break this barrier.

SAFFLOWER

Safflower (*Carthamus tinctorius L.*), a member of the family Compositae or Asteraceae, is a branching, thistle-like herbaceous winter annual plant, with numerous spines on leaves and bracts mainly grown in dry hot climates as an oilseed, birdseed or for its flowers, used as dye and for medicinal purposes.

World distribution and production: Traditionally, cultivation of safflower has been in a band from the Mediterranean to the Pacific Ocean at latitudes between 20°S and 40°N, wherever a hot, dry climate suits the crop. The crop is widely used for dye production in southern Germany and Alsace in France. As a minor crop, safflower is particularly vulnerable to the vagaries of the market, pests and adverse weather patterns. These have contributed to wide swings in production and seeded area in several countries.

Crossing techniques: Safflower is a predominantly self-pollinating crop, with the genetic potential for over 90% selfing, although environmental conditions may result in outcrossing up to 50%. A mass emasculation technique developed at the Nimbkar Agricultural Research Institute, Phaltan, takes lesser time than conventional emasculation of individual florets. On the initiation of first flower, 5–10 fully developed capitula from the top 4–5 branches of each plant are covered with low to medium density polythene bags. The hybrid programme in safflower was initiated by A.B. Hill in 1974 with cytoplasmic male sterility system. The average yield advantage of recent hybrids, compared at several sites in California, Arizona, North Dakota, Canada, Pakistan, Mexico and Spain, was 127% over the best parental line. In hybrid, average oil level of 34% in 1983, increased to 42% in 1994. Hybrids with 45% oil content are currently being developed.

Work at NARI

(i) Development of high yielding and high oil containing hybrids of safflower for minimal irrigation areas.

(i) Technology for the production of spiny and non-spiny safflower hybrids based on genetic male sterility system has also been developed.

(iii) Developed cytoplasmic male sterility systems in safflower through mutagenesis with streptomycin treatment and by interspecific crossing.

(iv) Studied genetics of flower yield and its components in safflower in order to develop hybrids producing high petal and seed yield.

Characteristics of Safflower Hybrids Released by NARI

S. No.	Cultivar	Year of release and recomm-ended region for production	Average seed yield (kg/ha)	Increase in seed yield over check (%)	Oil content (%)	Characteristics	Area under NARI cultivars (ha)
1.	NARI-NH-1 (First non-spiny hybrid in the world)	2001 All India	1936	10 A-1	35	Non-spiny, high oil containing, tolerant to foliar and wilt diseases.	1,500
2.	NARI-H-15 (Spiny hybrid)	2005 All India	2201	19 NARINH-1	30	Early maturing, tolerant to aphids and wilt.	—

The major findings of safflower research carried out under the AICRP during 2006–07 are as follows:

1. Release and notification of new spiny hybrid NARI-H-15
2. Dissemination of technology of seed production of non-spiny hybrid NARI-NH-1
3. Development of cytoplasmic male sterility system through interspecific crossing
4. Identification of thermo-sensitive genetic male sterility (TGMS) in safflower
5. Evaluation of thermo-sensitive genetic male sterility(TGMS)-based hybrids in safflower

CANOLA

Canola was developed in the early 1970's using traditional plant breeding techniques by Canadian plant breeders to remove the anti-nutritional components (erucic acid and glucosinolates) from rapeseed to assure its safety for human and animal consumption. The oil of canola is very low (< 7%) in saturated fatty acid. In Canada there are three main groups of herbicide-resistant canola: Roundup Ready and Liberty Link varieties were produced by genetic modification, whereas Clearfield was developed through mutagenesis. The transgenic gene inserted into the canola plant for herbicide resistance is carried by a protein. During processing, removal of all the protein from oil made it as of oil from non GM. Another innovation is the development of hybrid canola varieties. Hybrid increases yield and acreage both. Hybrids are expected to be sown on 80 per cent of canola acreage due to introduction of Roundup Ready, In Vigor and Clearfield hybrids.

Just a decade ago the canola industry was a buzz with word of a new seed technology that had the potential to shatter even the highest open pollinated canola variety yields. The recent technology of hybrid canola would change the entire canola industry by yield and performance expectations. Creating a hybrid is more complicated with canola than other crops, because of hermaphrodite nature. However, cross pollination is a requirement of hybrid.

Canola hybrid is simply the result of cross-breeding two lines of canola. Research in greenhouse showed that hybridization between two distantly related lines of canola resulted in 45% higher yield than either of the parent. However, production of large quantity of hybrid seed by hand is economically impractical. Since *B. napus* varieties are mainly self-pollinated, the self-pollination of the parental lines must be controlled to make hybridization commercially feasible.

CMS system: To date, several approaches have been attempted to develop hybridization systems in *B. napus*. The first relatively successful programs utilized more traditional hybrid breeding methods such as cytoplasmic male sterility (CMS). Researchers discovered that some *Brassica* species and close relatives had male-sterile cytoplasm (material surrounding the nucleus of a cell). At the cellular level, fertility is controlled by an interaction between the nuclear and cytoplasmic genes. The CMS system for canola hybridization depends on the mutation in certain cytoplasmic bodies that failed to develop functional pollen. "A" line can not be self pollinated due to sterile cytoplasm. Therefore the A line is maintained by hybridization with isogenic B line. The restorer line contains nuclear genes that compensate for the defect in the cytoplasm and restore fertility to the F_1 hybrid with heterosis. The first commercial CMS *B. napus* hybrid, Hyola 40, was registered in 1989, followed by the very popular hybrid Hyola 401 in 1991.

PGS system: A novel hybridization system was developed by Plant Genetic Systems in Belgium through biotechnology. This system involves the use of two parental lines. The first parental line is male sterile and does not produce viable pollen grains and cannot self-pollinate. A gene isolated from a common soil bacterium and inserted into the parental line causes this nuclear male sterility. The gene controls production of a specific enzyme in a specific anther cell layer and at a specific stage of anther development resulting in no pollen production. The second parental line contains another gene, obtained from the same common soil bacterium that produces an inhibitor enzyme that counteracts the sterility enzyme in the first parental line thus restoring fertility. A gene that confers tolerance to the herbicide glufosinate ammonium was inserted into both parental lines. When the two lines are crossed the progeny is a 100% true hybrid, since fertility is restored, the hybrid plants are fully fertile and produce seed. The first herbicide-tolerant hybrids "3850 and 3880" were registered in 1996.

***Barnase/barstar* technology:** The whole of the process of male sterility and selective removal of undesired hybrids from the male fertile plants, is performed elegantly by the *barnase / barstar* gene system, in which herbicide resistance is linked with male fertility, so that the herbicide will kill the male fertile lines, leaving the seed producing sterile male plants unharmed. The system is used only to develop novel hybrids and the farmer is provided with highly viable seed of uniform quality that produces fully (male and female) fertile plants.

The whole set of gene systems used in canola and mustard are a complicated but an indigenous development in rDNA technology that confers nuclear male sterility to self-pollinating plants in a stable manner to produce hybrids using a female restorer line. It helped to produce improved hybrids of such crucifer crops as canola and mustard, which is impossible without this gene system. The *barnase / barstar* gene system is meant to produce hybrids, and not herbicide resistant/tolerant crops.

Roundup Ready canola is resistant to Monsanto's herbicide Roundup and AgroEvo's GE canola to the Liberty. The same *barnase / barstar* gene system is being used in India to produce herbicide resistant mustard hybrids.

***In Vigor* hybrid seed production:** The seed production of *In Vigor* takes place by seeding alternating bays of males and females. The male plants can produce pollen and are Liberty tolerant and the female plants contain the male sterility gene which is also linked to the Liberty gene. As a result, the herbicide Liberty can be applied to the entire crop to keep the field weed free. The application of Liberty also prevents production of pollen in female line with exceptional level of hybridity. "The effort and resources required to go from an isolated seed line to a bag of In Vigor hybrid canola seed are a testament to Bayer Crop Science's commitment to product purity and excellence," says Lamont.

Assessing the potential for hybrids to contribute to overall feedstock: Perhaps the closest precedent to the development of canola hybrids was the development of corn hybrids in the later half of the 20th century. The development of hybrid technology in this crop led to yield increase that averaged between 1.52 and 2.85 bu/acre per year over the years from 1950 through 2000, and a quadrupling of average yields over 80 years since the technology gained wide use. It is reasonable to assess the progress of canola to determine whether the innovation has the capacity to drive yields in canola as it did in corn.

Hybrid canola was grown in Canada in the early 1990's with the introduction of Hyola 401. By the late 1990s hybrids became more popular as growers adopted glufosinate tolerant hybrid technology. Over the past decade, numerous other hybrids have been introduced into the market. Hybrid canola out-yielded open pollinated varieties in all the three soil zones of long, mid and short season zones across the prairies by 15 per cent or more. During the late 1990's and early years of 2000, continued progress in hybrid technology widened the gap between the new technology and open pollinated varieties.

- Growers have capitalized on the genetic potential of hybrids and have begun to realize real yield gains.
- Hybrids, introduction of high stability canola oil, global biodiesel demand, were responsible for our success.

CONCLUSION

India's seed industry has grown in size and level of performance over the past four decades. Both private and public sector companies/corporations are involved with the production of seed. The public sector component comprises two central corporations, *viz.* National Seeds Corporation (NSC) and State Farm Corporation of India (SFCI) and 13 State Seed Corporations. The private sector comprises around 150 seed companies, which include national and multinational companies and other seed producing/selling companies. Seed component in 1990–91 comprised proprietary hybrids, public-bred hybrids and open-pollinated varieties. In terms of quantity and value, seed of open-pollinated varieties was the largest followed in order by public hybrids and proprietary hybrids. Seed of the proprietary hybrids formed a significant portion of the total seed market. In terms of quantity, although proprietary hybrids had only 32.34% share of the market, in terms of value their share was 76%. In contrast to the 1991 scenario, the estimates in 1998–99 presented a different picture with proprietary hybrids growing at the expense of the public hybrids. The area planted to the bought seed increased by 3% over that of 1990–91 and the market size expanded significantly in terms of quantity and value. The total market for purchased seed was 8.64 lakh tons valued at Rs 2,249 crores. The component of proprietary hybrid seed was estimated to be around 51,314 tonnes in 1998–99 as against 19,300 tonnes in 1990–91 and valued at Rs 636 crores as against Rs 95 crores, respectively. The volume of public bred hybrids came down to 38,704 tons in 1998–99 as against 59,671 tons in 1990–91. The volume of seed of open pollinated varieties (OPV) increased by 51% to

774,881 tonnes.

Hybrid rice has been proved to be a very effective approach which greatly increased yield not only in China, but also outside China. Vietnam and India have commercialized hybrid rice for years. Last year about 600,000 hectares were covered with rice hybrids in Vietnam. On an average, the yield of rice hybrids is 6.3 t/ha while that of the inbred varieties is 4.5 t/ha. Vietnam becomes the second largest rice exporting country in Asia by planting of hybrid rice on large scale. Besides, many other countries, such as the Philippines, Bangladesh, Indonesia, Pakistan, Ecuador, Guineas and the USA have also achieved great progress in extending hybrid rice technology. Recently, a number of experimental trials and large scale demonstrations in farmers' field conducted in these countries have shown that hybrid rice can significantly outyield the pure line varieties. The above facts clearly indicate that hybrid rice technology first developed by China is also effective in greatly increasing rice yields worldwide.

The ever-forwarding technology improvement by scientists, its dissemination by seed industry by extension workers and the policy and financial support by national and local governments contribute greatly to the success in the development and use of hybrid rice technology. Relying on scientific and technological advances and the efforts from all other aspects, particularly from FAO and IRRI, it will have a very good prospect for commercial production of hybrid rice and continue to play a key role in ensuring the future food security worldwide in the new century.

Progress made in large scale seed production in India, in the initial phase, has been impressive as compared to the progress made in China at similar stage. In China, it took more than five years to attain an average seed yield of 1.0 t/ha and only after a decade the average seed yields of 1.5 t/ha were attained. In India, the average seed yields of 1.5 t/ha are being obtained after five years of initiating large scale seed production. Seed yields can be increased further by utilizing the new CMS lines with higher outcrossing potential, now under development in Indian hybrid rice breeding programmes and through further refinement of seed production technology.

Hybrid rice is likely to become popular and adopted in large acreage during the next decade. Huge quantities of hybrid rice seed will be needed. Thus there is ample scope for all the private and public sector seed agencies in India to produce large quantities of hybrid seed in the years ahead to popularize rice hybrids.

The development of improved CHA technology of the past decade has allowed breeders of hybrid wheat to avoid many of the problems involved with the CMS system. Large numbers of hybrids can be produced with breeding lines or populations of interest, and heterotic population structures can be developed and improved. Hybrid seed yields, economically important for the survival of the hybrid seed industry, can also be evaluated with the use of CHAs. The heritability of traits important in improving the cross-pollination required in hybrid seed production are moderate to high, and thus reasonable genetic progress can be expected in developing parents with improved seed yield efficiency. The opportunity and possibility of developing economically acceptable hybrids that appropriately reward the seed company and the farmer, and provide an increase in the world food production has considerably improved. Hybrids produced with CHA technology will continue to enter official trials and from there proceed to the marketplace. The CMS based hybrids will have a lot of scope to enhance wheat production and to play a major role for food security.

In the last 50 years, the average global productivity (yield ha^{-1}) of pigeonpea has shown little improvement. The advent of hybrid technology has given hope. The extensive testing of GMS based hybrids and limited evaluation of CMS based hybrids have demonstrated that high levels of exploitable heterosis are present in pigeonpea. Since an excellent CMS system is now available and the level of natural outcrossing is sufficient to produce large quantities of hybrid seed (Saxena, 2006), the prospects for commercialization of pigeonpea hybrids are bright. The hybrid pigeonpea seed production programme have now been taken up by a number of private seed companies in India and it is expected that the pigeonpea hybrids will be available for general cultivation in 3–4 years. To maximize the benefits from this technology more research and development efforts are needed in the near future. The *Helicoverpa* pod borer remains number one enemy of pigeonpea; therefore our next target should be to develop high yielding hybrids incorporating Bt gene(s) and good nutritional quality. Madhya Pradesh has been identified as a potential location for hybrid seed production of ICPL 2671. The performance of this hybrid is also excellent in terms of productivity in the central India. Hence it could be concluded that hybrid technologies can add to the pulse production and nutritional security in this country.

Hybrids in safflower developed at NARI have become popular and it is likely that more hybrids will be commercialized in near future and bring nutritional security in India.

Innovations in canola, through the introduction of herbicide tolerance, have reduced farmers' costs and the use of natural resources in its production. This technology also finds a promise in the mustard growing areas in the country.

The transgenic gene inserted into the canola plant to produce Roundup Ready and Liberty Link In Vigor varieties herbicide resistance with a protein. All protein is removed from canola oil during processing, so produces canola oil contains no GM material and is identical to canola oil from a non-GM canola plant. Another innovation is the development of hybrid canola varieties. Hybrids give higher yields and are increasing in acreage. New Roundup Ready, In Vigor and Clearfield hybrids have been introduced in cultivation and hybrids are expected to occupy 80 per cent of canola acreage in 2008. These technologies have not been introduced in this country till date. There has been a definite advantage of these technologies in fulfiling the requirement of global oil production. Hence it could be concluded that lot of efforts are needed to promote the emerging hybrid technologies with huge public investments under NARS and promote of these technologies through seed extension support systems under the developmental agencies.

REFERENCES

Briggle, L.W. 1963. Heterosis in wheat—a review. *Crop Sci.*, **3**: 407–11.

Ma, A.Q. & Sorrells, M.E. 1995. Genetic analysis of fertility restoration in wheat using restriction fragment length polymorphisms. *Crop Sci.*, **35**: 1137–43.

Pickett, A.A. 1993. Hybrid wheat: results and problems. *J. Plant Breed.*, **15**: 1–259.

Saxena, K.B. 2006. Seed production systems in pigeonpea. ICRISAT, India.

Virmani, S.S. and Edwards, I.B. 1983. Current status and future prospects for breeding hybrid rice and wheat. *Adv. Agron.*, **36**: 145–214.

Wilson, J.A. and Ross, W.M. 1962. Male sterility interaction of the *Triticum aestivum* nucleus and *Triticum timopheevii* cytoplasm. *Wheat Info. Serv.* (Kyoto), **14**: 29–30.

Wilson, J.A. 1968. Problems in hybrid wheat breeding. *Euphytica Suppl.*, **1**: 13–33.

Wilson, J.A. 1984. Hybrid wheat breeding and commercial seed development. *Plant Breed. Rev.*, **2**: 303–19.

14 Public-Private Sector Partnership

Rajesh Mishra*

ABSTRACT

Some success stories, existing scenario and need of the hour for partnership between Public and Private Sectors are presented.

Key words: Commercialization, marketing, public-private partnership

The public-private sector partnerships existed for many years in the world for various fields and have proved successful. In India the zeal for such partnership has been recent, it is a healthy sign.

Some Success Stories

In order to benefit from the past and current experiences a few examples can serve as an indicator of partnership.

- The first is an example from the US and is particularly important because it highlights a proven and successful example.
- National Centre for Agricultural Utilization Research (NCAUR) is the largest of four USDA-ARS federal utilization centers and a world-class bioscience research facility. Its mission is to find new uses and market for US agricultural commodities. The investment in NCAUR basic and applied research program has resulted in hundreds of new products and technologies benefiting the US agricultural industry.
- Biotechnology Research Development Corporation (BRDC) is privately owned and had seven corporations as partners. There purpose is to discover and fund technology that holds promise to make a significant impact in the international market.
- In 1977, scientists at NCAUR began a research project to incorporate starches and other plant derived polysaccharides into petroleum based polymers with the objectives of reducing prices of packaging materials

* GM (R&D) Seeds, Dhanuka Agritech Ltd., Hyderabad.

and reducing environmental hazards from non-biodegradable polymers. Although they had some success the products developed had less then the desired mechanical strength and they were not biodegradable. However their research builds knowledge based on starch and other polysaccharides chemistry.

- Dow Chemical Company (a member of BRDC) developed a family of synthetic polyesters that were completely biodegradable and compostable and had superior mechanical strength also. But their cost was prohibitive. Dow tried to blend their products with inexpensive fillers but that led to an undesirable product.
- Dow went to BDRC who then got in touch with NCAUR. Within one month BDRC and ARS were working on a research project and within two months BDRC began to file a family of patents to cover the intellectual property that was developed out of that cooperative research endeavour. Dow scientists worked very closely with the ARS scientists. By 1997, Dow had taken an exclusive license to the technology of producing loose fill packaging material which represented about 150 million cubic feet of polystyrene every year.
- Thus this government-private sector partnership has been beneficial not only to the partners but also to the consumers and environment.
- The second example is that of ICRISAT at Hyderabad which has formed a Consortium to develop links with the private sector. This partnership allows private sector seed companies to access the vast germplasm pool of ICRISAT. The benefits of such a partnership will be evident in the near-term in the form of new hybrids in various crops.
- The third example—largely a private sector driven initiative is e-Chaupal, where in an Indian corporate (ITC) has taken the lead in developing end-to-end solution in selected villages in some states. These villages are connected through computers and the villagers are provided information and guidance on not only the agronomic management of their crop but also on the marketing of produce. Most of the state governments are not only appreciative but also partners in this venture by providing support.

Existing Scenario

Agriculture education, research and extension form the holy trinity of a well-founded agricultural system and most nations have many state funded institutions performing these roles. In India, education and research are generally done through the ICAR-SAU system while extension is the responsibility of the state departments of agriculture.

Table 14.1. Various activities and agencies in agriculture

Activity	Agency
• Agriculture	the public
• Education	mainly done by the SAUs
• R&D	ICAR institutes, SAUs
• Regulation	various Central and State ministries
• Extension	state Dept. of Agril in conjunction with SAUs
• Seed production/supply/quality	state agencies and corporations

In our system there is no dearth of aspiring students, scientists and extension personnel, however it is evident that it is becoming difficult for the Central and State Governments to fund the system through their own resources. The main problem is that most of the funds are utilized in paying staff salaries with very little left for research projects.

Defining the public partners: Currently any institution funded by public money would fall into this category. It could be ICAR or any of its institutes like IARI, a SAU, a public bank, any ministry, department or agency of the various states, or a member of CGIAR system. The list is endless.

Some of the other relevant questions that need to be answered are:

- Who are the partners?
- Are they empowered to enter in partnerships?
- What is on offer?
- What could be the mode or models of partnership?
- Is the public sector partner ready to partner in letter and sprit?

Despite many examples of successful partnerships there may be others that really have not worked out for various reasons, but it is encouraging to note that the mood is upbeat and the Central Government and various State Governments are developing a favourable attitude.

Partners in research: As far as research and development is concerned, it is abundantly clear that as a result of cost constraints, new technologies are quite out of reach of small and mid-sized companies.

Over the past one year ICAR has been expressing willingness for partnership with the private sector, they have worked out the guidelines for partnership. This is a welcome step as guidelines will be helpful in developing partnership. Various SAUs have started action for developing partnership as per the guidelines.

The broad partnership will be in the form of technology development at public sector institutes and their commercialization is best left to the private companies.

The nature of R&D requirement of private companies has been limited to that for varieties with better agronomic, pest, insect or disease resistance traits, the quality component will be an added advantage. The source of such varieties has been either local research institute or foreign breeders and companies. Most Indian seed companies have spent a limited amount of their resources on R&D. Recently, because of a change in the nature of research requirements from agronomic to biotech, many medium and big companies have increased there budgets and activities.

In case of public research the past focus has been on open-pollinated varieties and is generally spread over a large number of species. Given the large number of Universities the research tends to be repetitive with the same aspect being researched by many researchers. Most SAUs have large infrastructures but limited funds to run their research programs. Also a limited or no interaction with the private seed companies has been witnessed. Most of the research has been on agronomic aspects and interaction has been limited to those with fertilizer, pesticide or agricultural equipment companies.

In the changing scenario there are several areas in which the public institutions and private companies can work as partners, some of which are summarized in Table 14.2.

Table 14.2. Areas of cooperation

- Germplasm
- Biotechnology and other technologies
- Research facilities-labs, land and personnel
- Research grants and fellowships, contractual research
- Training
- Foundation seed production
- Seed quality and food processing laboratories
- Multilocation trials

Though all areas are important, the primary need for any research partnership to succeed will involve the following aspects which private partners will expect to be non-negotiable.

The Need of the Hour in Research

Accuracy	Requires experience and accurate machines, spares, back-up
Zero	Tolerance for inaccurate results
Speed	Short time line
Economy	Cost effective
Single window	Most services under one roof

Confidentiality	Utmost secrecy
Liability	Opportunity loss in case of failure to deliver results.

Ensuring that all the above parameters are met satisfactorily will be important in determining not only the number of partnerships but also the probabilities of there success. Given the optimism in favour of getting into a P3 relationship, most public sector institutions will possibly want to partner with a private company but the emphasis has to be on partnerships that can deliver some real value. A good strategy will be to focus only on those areas of research that can fulfil all the above criteria.

One way forward could be to designate ICAR institutes/SAUs to conduct research on the areas of their competence. Some funds for such research can be provided by individual firms or associations. Such institutions also need to have nominees of seed associations on their research committees. This will help to make these institutes more market oriented. Research in such institutes needs to be of international standards for them to attract the interest of companies.

The SAUs and there colleges with vast acreages of land could focus their efforts on agronomic, plant protection and breeding research as also on conducting multilocation trials, seed production and on developing food processing technologies. These institutions could perhaps be discouraged from spending scarce resources on hi-tech research which requires a much higher level of funding.

Seed production and supply: The PPPs are not only limited to research. One important area is that of seed production and supply. It is quite clear that substantial amounts of open pollinated seeds are planted in India and those SRRs are very low, with an average of 30%. Many private companies are not very keen to get into the open pollinated seed business because of the subsidy structures that makes it difficult even a marginal profit. That leaves this segment open either for the public sector companies or some private seed traders. But if the SRRs have to go up in open pollinated it is imperative that the private seed companies should be also encouraged.

SAUs with their state-wide research and production farms are ideally suited for conducting multilocational trials. They can also be partners with private companies for production of breeder or foundation seeds. With the new seed laws the SAUs will also be able to conduct DUS or VCU tests.

SAUs can also be given the mandate of conducting the seed testing, currently conducted by state seed quality laboratories. This will not only reduce the burden of the DoA but also speed up matters. SAU seed science departments are generally more modern as the quality tests and procedures are concerned.

Hybrid developed at public sector: It is true that many hybrids bred through public sector research still lying in their shelves. The reason may be many, but in high volume crops the main constraint is seed production. Most of the public sector institutions lack in the facilities for seed production locations for the hybrids.

The investment made through public funding does not reach to the end user, *i.e.* farmer. To address this situation there is a urgent need to hand over these hybrids through MoU to private sector so that they can reach the farmers.

In many crops, especially low volume vegetable crops the situation is reverse. Good hybrids are developed by the private sector but sometimes the seed production faces constraints. Here public sector institutes can be helpful, and for proper profit sharing seed production model needs to be developed.

Commercialization of public sector hybrid: Hybrids developed in pubic sector need to be passed on to the private sector for marketing. Since private sector has expertise in production and marketing the spread of hybrids will be fast, the revenue generated by the private sector will be shared between the partners and will be wise that the public sector reinvest the earnings to augment research for developing new products.

The private sector when interested in marketing right of any public bred hybrid, they have following considerations in there mind

- Continuous supply of parental lines as per the demand
- Hundred per cent genetic and physical purity of the parental lines
- Benefit-cost ratio.

These aspects are very important, as marketing of any hybrid requires huge investment. The hybrid needs to be tested in multilocations and at the location of better performance large scale demonstrations must be organized. These activities take at least two years and sizable investment; therefore the private companies look for a MoUs of long duration only so that they can recover the cost of investment.

Once they are sure about the performance of the hybrid the next step will involve seed production as per demand and for that a continuous supply of parental lines form the public sector is essential, it is advisable that a rolling plan is prepared by the partners in advance so that there is no miss-mach between demand and supply.

The next important point relates to the purity of parental lines. If the parental lines are either genetically or physically impure it will result in production of impure hybrid seed. The farmers, farmers associations and NGOs are highly sensitized and if such mishaps occur, the seed setting happen company will have to pay huge compassion. To address this situation joint monitoring of seed production must be undertaken.

A seed production team from the public sector who offered there hybrids to private partners need to be equipped with most updated knowledge of seed production of the parental lines. if required, the private sector partner can impart training or made arrangement for training to the partner.

The cost: Benefit ratio is most important for any private sector seed company, including the cost of parental line, royalty etc. If the cost is too high then they will loose profit margins. Therefore a win-win formula for all collaborators needs to be developed.

Subject Index

A

Agribusiness
- Consultancy services 207
- Credit 206
- Crop diversification 206
- Export promotion 205
- Government initiatives 205
- Information technology 205
- Initiatives 206
- Organizations 204
- Problems 204
- Prospects 204
- Venture capital scheme 208

Alternate locations 214
Alternative areas 8
Anticipated benefits 11

B

Buffer stocking 9

C

CGMS 198
Commercialization 250
Contingency planning 9
Cost: benefit ratio 250
Credit of IPRs 211

D

Demand assessment 199

E

Economics of hybrid rice seed production 215
Economics of seed production 11
Emerging issues 6
Environmental implications 3

F

First generation hybrids 209

G

Genetic purity tests 12

H

Heterosis 197
Hybrid canola
- Anti-nutritional components 237
- *Barnase / barstar* technology 239
- Glufosinate 239
- *In Vigor* hybrid seed production 239
- Liberty gene 239
- PGS system 238

Hybrid castor
- Achievements 146
- Future strategies 149
- Maintenance 146
- Seed production 147

Hybrid maize
- Advantages 42
- Baby corn 55
- Biological value 62
- Breeding cycle 52
- Composites 52, 54, 55
- Crop inspection 47
- Detasseling 47
- DNA-MAS 62

Double cross 42, 55
Ear rots 62
Early maturity 54
Earthing up 47
Economics 48
Flint lines 55
Gene pools 56
Genotyping 63
Germplasm 56
Heterotic Pools 56
Insect pests 46
Isolation distance 44
Leucine- isoleucine 61
Lysine 61
Male : female ratio 44
Marker aided selection 63
Marker assisted selection 52
Maydis leaf blights 53
Medium maturity 53
Microsatellite markers 62
Modified mass selection 55
Narrow-based seed parents 60
Nutrient management 45
Off-type 47
Opaque2 62
Phenotypic selection 62
Pollen parent 43
Popcorn 55
Prerequisites 43
Processing 48
Quality protein maize 52, 61
Recovered lines 55
Release 55
Seed parent 43
Seed rate 45
Seed replacement 41
Seed treatment 45
Seed village 42
Shelling 48
Single cross hybrid 48
Specialty corn 55
SSR markers 62
Sweet corn 55
Synchrony 42, 44
Three-way cross 42
Triple Cross 55
Tryptophan 61
Turcicum leaf blight 53
Water management 46
Weed management 46

Hybrid pearl millet
A_1 cms system 112
Alternative hybrid options 118
Benefit-cost ratio 137
cms systems 110, 113
Crop scenario 108
F_1 seed parents 111
Hybrid parents 120
Hybrids 109
Inbred pollen parents 113
Inbred seed parents 111
Inter-population hybrids 117
Linkages 133
Maintainer line 110, 120
Male-sterile line 110
NND_2 112
New markets 132
Nucleus seed production 119
Open-pollinated varieties 109
Population seed parents 112
Protection 136
Restorer line 110,113, 121
Seed class 119
Seed marketing 129
Seed processing 127
Seed production 123
Seed production management 125
Seed storage 129
Single-cross hybrids 116
Three-way cross hybrids 117
Top cross hybrids 116
Top cross pollen parents 113

Hybrid pigeonpea
Apis mellifera 156
C. scarabaeoides 157
CMS 233, 156
CMS system 157
Cost of hybrid 235
Diversification of parents 160
Equitable access 234
Fertility restorers 156

GMS 233
Hybrid Technology 232
Isolation 159
Isolation distance 235
Megachile species 156
Out-crossing 235, 156
Pollen transfer 156
Seed cost 160
Seed Production Technology 159
Short-duration hybrids 158
Sterility mosaic 161
Wilt 161
Hybrid rice
Achievements 15
Agro-technology 21
Amylose content 17
Anther culture 27
Anthesis 21
Apomixis 19
Aroma 17
Benefit- cost ratio 35
Biotic/abiotic stresses 18
Cheaper chemical composition 22
Cina 38
Constraints 36
Cost of seed production 22
Cultural management 24
Cytoplasmic male sterile line 23
Double transplanting 27
Economic viability 224
EGMS 19, 23
Flag leaf cutting 21
GA_3 19, 21,22
Genetic erosion 37
Genetic purity 20
Grain quality 17,37
Guidelines 23
Impact 35
Importance 14
Informal seed systems 31
Isolation 222
Isolation distance 22
Large scale seed production 20
Linkage 33
Low head rice recovery 37
Maintainer 23
Maintenance 27
Management strategies 19
Marketing 31
New areas 28
Non- synchronization 221
Non-aromatic CMS lines 17
One-line hybrids 19
Out-crossing rate 21
Paired crossing 27
Parental line improvement 20
PGMS 23
Plant Variety Protection and Farmers' Rights 32
Pollen load 21
Pollen parent 23
Poor grain quality 221
Popularization 30
Protection 34
Quality control 26, 224
Ratooning 26, 27
Regulatory mechanism 32
Restorer 23
Role of women 38
Rouging 23
S line 23
Scenario 13
Seed cost 37
Seed laws 32
Seed-borne diseases 37
Stickiness 17
Subsidies 33
Supplementary pollination 23, 25
Synchronization 19, 21
TGMS 23, 27
Timing of seeding of the parent lines 223
Two-line hybrids 19
Two-line system 23
Vegetative propagation 26
Yield advantage 37
Hybrid safflower
Cytoplasmic male sterility system 236
Genetic male sterility system 144

Heterotic potential 144
Hybrid seed production 145
Maintenance 144
NARI 236
Pair-wise selection 145

Hybrid sorghum
Alcohol industry 84
Bio-energy harvest 86
Cytoplasmic-genetic male sterility 72
Dairy industry 84
Durra caudatum 79
Economic viability 97
Export possibility 87
Fodder sorghum 83
Food-fodder 84
Forage sorghum 74
Grain mold 73
Guinea caudatum 79
Industrial sorghum 83
International scenario 69
Kafir milo 76
Kafir Shallu 77
Kharif sorghum 72, 83
Ms lines 73
National scenario 70
National seed policy 68
Poultry industry 84
Profitability 96
Rabi sorghum 74, 82
Seed production 96
Seed supply system 68, 91
Starch industry 85
Striga resistance 79
Sweet stalked sorghum 75
Value-addition 75
Variety improvement 79
Zera zera 73

Hybrid sunflower
Catch crop 140
Cytoplasmic male sterility 140
Helianthus petiolaris 140
Parental lines 141
Photoperiod -insensitivity 140
Seed production 142
Thermo-insensitivity 140

Hybrid wheat
Aegilops caudata 167, 227
Alloplasmic wheats 176
Breeding scheme 176
Buitre lines 172
CHA 174
Chemically induced male sterility 177
Chromosome 4A 180
CIMMYT185
CMS 167
Commercial feasibility 169
Cornerstone male sterile 180
Cytoplasmic sterility 227
Effect of cytoplasm 176
EMS 182
Fertility restoration system 175
Fi genes 176
Floral biology 167, 172
Future strategies 187
Gametocide 227
Genetics of male sterility-fertility system 175
GMS 181
Heterosis 169
Indian effort
Industrial use 199
Intellectual Property Rights 2
Inter- project linkages 12
Introgressing genes 172
Issues and strategies 7
K-type 176
Licensee 213
Limiting factors 199
Maintenance of parental lines 214
Male sterility system 174
Marketing 215, 250
Mega zones 166
Micronutrient deficiency-induced male sterility 183
Neglected crops 10
New Policy on Seed Development 209
Out-crossing 172
Parental diversification 172

PCMS 182
PGMS 183
Probus male sterile 180
PTGMS 183
Pugsley's male sterile line 180
Rf genes 176
RFLP 229
S 738 181
Scope and importance 166
Seed Production 184
Sirocco male sterile 180
Stay green 172
Taigu 181
TGMS 183
Triticum timopheevii 167
UC 9109-9 181
VE system 180
VE 181
V-type 176
XYZ system 179
Yield components 171
Yuma' x 'Capeiti' 181

P

Plans & policies 6
Plant Breeders' Rights 11
Production of parental lines 214
Public – Private partnerships 216
Seed quality enforcement 12
Seed replacement rate 199
Seed standards 12
Sharing of breeding material 211
Skilled manpower 10
Strategy 5
Success stories 245
Supply management systems 10
Trade liberalization 201